THE PSYCHOLOGY OF RISK TAKING BEHAVIOR

ADVANCES IN PSYCHOLOGY

107

Editors:

G. E. STELMACH

P. A. VROON

NORTH-HOLLAND
AMSTERDAM • LONDON • NEW YORK • TOKYO

THE PSYCHOLOGY OF RISK TAKING BEHAVIOR

Rüdiger M. TRIMPOP

Department of Industrial/Organizational Psychology
Ruhr University Bochum
Bochum
Germany

1994

NORTH-HOLLAND
AMSTERDAM • LONDON • NEW YORK • TOKYO

NORTH-HOLLAND
ELSEVIER SCIENCE B.V.
Sara Burgerhartstraat 25
P.O. Box 211, 1000 AE Amsterdam, The Netherlands

ISBN: 0 444 89961 8

Transferred to digital printing 2007

Für Ilona Patricia

und

Leonie Franziska

Foreword

Habent sua fata libelli.

If the significance of a book is measured in terms of the amount of debate that is triggered by its appearance, "The Psychology of Risk Taking Behavior" will quite likely prove to be a significant publication.

One reason for making this forecast is that, in his attempt to explain risk taking in the human condition, Dr. Trimpop has been comprehensive as well as eclectic. Few authors have juxtaposed so many approaches and studies in the domain in question. While, on the one hand, this book seems to have achieved a nearly encyclopedic coverage of the relevant literature, the author has taken a selective approach in the process of developing theoretical views of his own. So, debate will be provoked by the fact that some readers will be delighted to see at least some of their favoured views represented in Dr. Trimpop`s theory, while others - having their own preferences - may be dismayed by the fact that theirs are not.

The two most debatable points are possibly the endorsement of risk homeostasis theory and the notion that people take risk for the sake of risk. Although the data base in support of risk homeostasis has grown significantly over the last several years and its opposition dwindled, the concept that risk taking is homeostatically controlled still constitutes a major source of controversy, even if this is mainly due to misunderstanding, wilful or innocent.

The "Psychology of Risk Taking Behavior" stresses that there are more justifications of risk taking than merely maximising economic utility. In the same vein and years ago, the theory of risk homeostasis has referred to the need to satisfy curiosity, to maintain an optimal level of psychophysiological arousal, and to the "dare-devil circus acts" committed by teenagers and young adults that may serve sensible mating choice and biological fitness for future generations. Risk homeostasis theory, however, never stated that people take risk for the sake of risk. On the contrary, it views risk taking as an inevitable consequence of purposeful action to other goals - other than risk that is.

And here, Dr. Trimpop´s book goes well beyond the 'parent' theory of risk homeostasis put forward by the supervisor of his doctoral dissertation and in doing this he opens a new perspective. He argues and energetically substantiates his reasoning by fully exploiting the extant supply of relevant literature in psychology and biology. Human beings are seen as predisposed to taking risks. It is proposed that risk taking is *intrinsically* rewarding - and not just tolerated in the anticipation of reaching goals that can be reached only at the expense of taking a degree of risk. Instead of 'an inevitable discomfort', risk is construed as having a positive hedonic value, as pleasurable.

This is a refreshing change from the more traditional treatises of risk taking that see it in the perspective of utility maximisation, with utility usually being of the economic variety. The traditional texts usually treat risk taking on a rational, cognitive level only. These are often full of statistics of calculable objective risk and no less replete with studies that seem to show that people are rather inadequate decision-makers in the face of uncertainty. In contrast, Dr. Trimpop emphasises that risk taking may be warranted on emotional and biological grounds and that cognition is merely one of the relevant aspects.

Moreover, risk taking is seen as biologically driven, unconscious and genetically programmed. Here we encounter one more feature of this book that undoubtedly will prove provocative and will elicit debate. Darwin and Freud are made to meet: risk taking as necessary for evolution and risk taking as principally pleasurable. Another 'psychodynamic' aspect of Dr. Trimpop`s risk motivation theory is the notion that there is a counter agent: the desire for skill, mastery and control which acts to counterbalance the need for risk. Control is also viewed as being pursued on a non-conscious level and its acquisition as *intrinsically* gratifying. In short, here we are presented with two opposing forces in the intrapsychic arena: the thrust for risk and the drive for control; two pleasures principles pitted against one another.

Wouldn't you agree, reader, that in writing a foreword to a book containing themes like these, I am taking little risk in predicting that this book will get you and many other readers involved in considerable debate?

"Habent sua fata libelli".

Terence, the Roman playwright from Carthage, was right some twenty-one centuries ago: ***books have their adventures***. He spoke from experience as his work was met by a great variety of response. It will, no doubt, be very interesting to follow the fortunes of the book you are holding in your hands.

Gerald J.S. Wilde, Ph.D.,

Professor of Psychology,
Queen`s University, Kingston , Ontario, Canada.
January 1994

THE PSYCHOLOGY OF RISK TAKING BEHAVIOR

Preface

During the last 15 years I have consciously and often voluntarily engaged in many so called high-risk activities. I am, for example, a motorcyclist and SCUBA-diver, I like driving fast cars, engage in parachuting, fencing, contact sports, travelled to the Arctic and the African jungle in a tent, to Israel during the Lebanese war and have engaged in many social risk taking activities, such as acquiring a Ph.D. in a foreign country. Some not necessarily conscious risk taking behaviors can be found in my outspokenness, and my sometimes blunt honesty. In all these situations, I have tried to find reasons for my fascination with these activities and the accompanying emotions and risk experiences.

One reason may just be that I may be somewhat abnormal (thereby justifying my becoming a psychologist), but the people I have met in these "hobbies" (or you may call it my life), have had very different personalities. The parachutists ranged from the 16 year old school-girl to the 61 year old supreme court judge, the motorcyclists from the members of motorcycle gangs to 72 year old members of peace groups, some SCUBA-divers were overweight primary school-teachers and others were safety engineers. The individual differences between these people did not allow me to classify them as being risk-loving, abnormal psychopaths, or to put them in any kind of personality category.

Furthermore, individual styles and types of behavior within the same situation were as different as the background of the people. One person wanted to parachute in spite of a broken leg, while another person quit after receiving a slight bruise. Some people were joking and casual, others were silent, withdrawn and checked their gear up to 15 times before the jump. Similar observations hold true for the other activities. The only common element seemed to be some form of joy in these activities, due to the perceived benefits, such as gaining social prestige, or earning money, or simply enjoying the feeling of being very much alive and in control of one's destiny.

Being trained to distrust my own biased interpretation of myself, these people´s personalities and situations, I decided to approach the problem scientifically. The following questions arose, and formed the purpose and outline for my Ph.D. thesis and this book which arose from it in 1993:

- What is risk taking behavior and how do we perceive risks?
- What personality type engages in risk taking behavior? And is it mainly determined genetically or by situational characteristics?
- What motivates people to take risks, and what do they feel before, during and after the performance?
- How do we control and master possible negative consequences of risk taking behavior and how do we compensate for different levels of perceived risks?
- Is there a connection between different types of risk taking (financial, social, physical) and can we explain it in one "holistic" model ?
- What does ***risk motivation*** mean for our daily life and how can we modify it?

To find answers to these questions I started to review the literature on risk and risk taking. I quickly realized that risk is an important research topic in the areas of psychology, politics, economics, medicine, engineering, sports, sociology, law, and just about every other possible area of human interest. In addition, a probably understated estimate of the amount of journal articles published in the last 25 years revealed more than 10,000 articles on risk in the areas of psychology and sports alone.

From 1988-1990 I faced the same problem that Yates (1992) mentioned in his introduction, namely the absence of comprehensive reviews of risk taking behavior. Thus, I decided to gain a deeper insight into the topic through an intensive literature review. I found that many researchers had been writing about risk for decades and had invested an enormous amount of physical, cognitive and presumably emotional effort into their particular facet of interest about risk taking. Although different theories seemed to state the opposite causal relationships or conclusions, they provided equally convincing empirical support. One example for different approaches is the debate whether risk taking behavior is determined genetically as a personality trait or situationally in rational cost/benefit analyses.

Taking myself as the most detailed source of information I have access to, I came to the conclusions that many different assumptions are correct at one time or another. Sometimes I make purely rational choices about risks to take, for example when investing money for my pension plan, but generally I tend to take more risks than some people I know, for example by engaging in high-risk hobbies friends of mine would shy away from.

The conclusion was, therefore, to assume that risk is a multifaceted phenomenon and that maybe different theories discuss different facets of risk taking. Thus, it is also conceivable that a specific theory can be empirically supported, while another study may show an allegedly contradictive effect. Such inconsistencies, however, may be due to differences in situations, the type of risk taking, the mood the subjects were in, or even the personality pattern of subjects.

As most theoretical approaches only focus on a small aspect of risk taking, different results could be more indicative of different experimental conditions than of any particular theory being completely wrong. This conclusion forms the basis for this book. Instead of trying to show that one particular theory is wrong I decided to take empirically well supported elements out of many different approaches towards risk taking and combine them to develop a holistic model of risk taking that allows for the multifacetedness of the concept of risk. Little did I know that 5 years after I had begun, Yates (1992) and Elander et al. (1993) still required researchers to formulate a risk theory with exactly these features.

I shall argue that risk taking is an inevitable behavior in any environment that changes and thereby presents some degree of uncertainty about the future. I shall present support for the postulate that risk taking has been and is an essential behavior in human evolution, and that it therefore possesses most likely a physiological and genetic predisposition, and an intrinsic reward mechanism.

A physiological predisposition, as manifested in risk-related personalities, however, is not the only, or even most important aspect of risk taking, as will be shown by studies which prove the importance of situational variables, both in shaping and rewarding risk taking behavior. I shall further argue that risk taking depends on the experience of risk, namely on physiological reactions, emotional experiences and cognitive assessments, as well as their interactive feedback loop. Thus, I postulate risk taking to be a concept combining individual, situational, physiological, emotional and cognitive factors, as well as thoughts and actions, each of which in itself is insufficient to describe the complete concept of risk taking accurately.

The outline of the book follows the research questions presented above.

In chapter 1, after having presented the purpose and outline of the book, I shall briefly discuss the history of research on risk. Then, I shall discuss different definitions of risk taking and provide my own definition in trying to answer the question: ***What is risk taking behavior ?***

In chapter 2, the concepts of risk assessment, risk perception, and risk acceptance across different people and cultures will be discussed to answer the question: ***How do we perceive risks ?***

In chapter 3, I shall discuss biologically based arousal and personality theories of risk taking. An intensive review of the literature on genetic components, personality models, evolutionary approaches with empirical results and implications for almost all types of risk taking behavior are discussed here to answer the question: ***Who engages in risk taking behavior?***

Chapter 4 will deal with theories about situational variables that are analyzed in utility theories and cost/benefit analyses to answer the question: ***Which situational factors influence risk taking behavior ?***

Chapter 5 will deal with emotional, and motivational aspects of risk taking to answer the question:
How is risk taking motivated and emotionally experienced?

Chapter 6 relates to the necessary counteragent of an intrinsically and extrinsically motivated risk motivation, namely the desire for control and mastery to answer the question: ***What controls risk taking behavior ?***

In chapter 7 I shall discuss the role risk taking behavior plays in the causation of accidents in traffic, sports and industry. Also, theories and empirical studies of accident prevention will be discussed to answer the question: ***What are the consequences of risk taking behavior ?***

Chapter 8 aims at understanding the behavioral adaptation, compensation and homeostatically controlled processes. Risk Homeostasis Theory is discussed with empirical and theoretical studies to answer the question: ***How do we adapt to the desire for and the control of risk taking ?***

In chapter 9 I shall combine elements from all of the above mentioned theories and empirical findings. Based on that I shall introduce Risk Motivation Theory (RMT) as an answer to the question:
Can risk taking behavior be explained holistically?

The final chapter 10 will summarize the ideas and findings presented before and draws conclusions to answer the question:
What are the implications of a risk motivation for our daily life ?

I hope that this book will offer the reader some interesting thoughts and a comprehensive review of this fascinating topic. Looking back to a number of positive associations with risk taking activities in sports and social life, and as a student, of course, also in my so-called financial life, I hope to offer a different point of view than many other researchers do, who seem to approach the topic from a purely rational, risk avoiding or merely "risk-tolerating" vantage point.

Furthermore, as I have received extensive education in clinical and physiological psychology, my focus is more on the intrinsic processes, many of which happen automatically and sub-consciously. Most other approaches, including the book by Yates (1992) do treat risk taking behavior from a cognitive, rational, decision making point of view. Although I definitely support the importance of those facets, I strongly believe and know that risk taking behavior in all of our lives is dominated by our feelings, our genetic heritage and of course the positive and negative experiences we have had with it throughout our lives.

Nothing ventured nothing gained is a proverb common to most languages and thus I hope to take you on a little adventurous journey to go where few psychologists have gone before.

Acknowledgements

A book like this can hardly be written without numerous sources of financial, informational and - very importantly - emotional support. The financial support came from the Ontario Ministry of Transportation, Canada and partly from my parents, Kurt and Margarete Trimpop.

The informational support came from Prof. Dr. Wiesenthal, Prof Dr. Cooper, Earl Briard from the IAPA, Prof. Dr. Julian Barling, and of course, Prof. Dr. G.J.S.Wilde. Further assistance in the formulation, layout and editing of the book was provided by Dr. Karyl MacEwen and Lars Adolph. All of these friendly folks mixed the information with positive feelings towards me, which makes it so much easier to utilize criticism for improvement and personal growth.

The emotional support was often, but not always mixed with helpful, constructive criticism. The major source for that type of support comes from my wife, friend, and colleague, Dr. Ilona Patricia Kryl. Finally, the last months of finishing this book were inseparably intertwined with the birth and early development of my daughter Leonie Franziska Trimpop. I did notice significant changes in my general attitude toward risk taking.

A German "barde" put into words what describes her type of support best:

I was never as productive, until you came to disturb!!!

Thanks to all of you!!

Chapter 4: Which Situational Factors Influence Risk Taking Behavior ?

Chapter 5: How Is Risk Taking Motivated and Emotionally Experienced?

List of Tables and Figures

Chapter 1: What Is Risk Taking Behavior?

1.1 A Short Description of the Long History of Risk Taking

The concept of risk has been a concern of human beings from the earliest days of recorded history and most likely even before that (Grier 1980; 1981). The Asipu in Mesopotamia dealt with predicting risks and risk management as early as 3200 B.C. Their approach was to identify important dimensions of the problem, design alternative actions, and collect data on likely outcomes (e.g. profit or loss, success or failure). They would then mark the different options with a plus or a minus sign, depending on how favorable the options were, and would finally issue a report to the client, etched upon a clay tablet (Oppenheim, 1977).

At the same time, farmers in Mesopotamia managed risks by taking out insurance on their production and loaning shares of their excess production for interest rates in return. Interest rates from 0% for friends to 33% for strangers reflect the perceived riskiness of loans (Grier 1981). Similar regulations with interest rates of up to 200% were reported from ancient Babylonia for merchants transporting goods through dangerous seas or bandit infested areas (Covello and Mumpower 1985).

Early recordings of risk analysis and risk management are found in the ***"Code of Hammurabi"***, issued at about 1950 B.C. Apart from several doctrines on laws and informal regulations for human interaction, the Code of Hammurabi formulated "bottomry contracts" for vehicles and cargo. These contracts contained the elements:

1) the loan on the vessel, cargo and freight,
2) an interest rate, and

3) a risk premium for the chance of loss and subsequent cancellation of debts (Covello & Mumpower, 1985).

By 750 B.C. bottomry contracts were widespread, especially in Greece, and according to Covello and Mumpower (1985) almost every voyage was covered with risk premiums of 10-25%. The concept of "general average" was established so that the risk premium was proportional to the risk of the journey, and owners would share the costs of damage to one of them collectively. With the Roman Empire taking over Western civilization, insurance and risk management contracts and even a rudimentary life and health insurance called "collegia" became institutionalized.

After the fall of the Roman Empire around 700 A.D. insurance almost disappeared and only reappeared in the 12-14th century, in form of marine insurance by groups of merchants in Germany, Italy, Spain and France. The Hanseatic League and Lombards established the Laws of Visby around the year 1300, in Spain the Ordinance of Barcelona was established in 1435, and around 1600 the Guidon de la Mer followed. In 1688 Lloyd's was founded in London, which later became the center of many areas of insurance. The moral issue of how much money can be placed on a human life was hotly debated, and is still an issue in cost/benefit analyses (Covello & Mumpower, 1985).

Other ancient risk management techniques included concepts of "strict liability", as found in the Code of Hammurabi (Hammer, 1980), or direct government interventions in the 5th century B.C. China, to change the "risk management technique" of sacrificing young maidens to reduce the risk of floods (Clark, 1980).

All ancient civilizations left records about preventive measures (e.g. building and fire codes) against major problems and risks they were threatened with, such as natural disasters, epidemics, pollution, food contamination, transportation accidents and occupational injuries (Hughs, 1975, cited in Covello and Mumpower, 1985). It can be seen that the addressed problems are still of major importance today, as many as 4000 years after they were originally identified and regarded as problems.

All these risk management attempts, however, were lacking a scientific risk analytic basis. One of the more sophisticated early analyses was performed by Arnobius the Elder in the 4th century A.D. (Grier, 1981). In his work "Against the Pagans" Arnobius thoroughly discussed risks and uncertainties of being a Christian or a pagan and employed a 2 x 2 matrix to represent the two alternatives of Christian versus Pagan on one hand, and uncertainties as to the existence versus the non-existence of God on the other hand. This type of risk analysis has been used frequently since and is called "dominance principle", a heuristic for decision making under risk and uncertainty (Grier, 1981). In 1657, when Pascal introduced probability theory, he extended Arnobius' matrix by inserting probabilities for the existence of God (Ore, 1960). With probabilistic analyses, risk assessments took an important step forward, as now uncertainties could be labelled with a specific probabilistic value.

Lightman (1983) named these steps in the development of quantitative risk analyses:

1) In 1692, Arbuthnot argued that different probabilities of potential causes of an event could be calculated.
2) Halley, in 1693, proposed very well developed life expectancy tables. He showed that popular assumptions about the effect of the moon on health and death was clearly wrong.
3) In 1728, Hutchinson examined the tradeoff between damage and utility in risky choice situations.

4) In 1792, La Place developed the prototype of modern quantitative risk assessment in calculating the probability of death with and without smallpox vaccination (Lightman, 1983).

In the 19th century, Von Bortkiewicz used data collected over a period of 10 years, and, building upon work by Poisson, he estimated the theoretical number of Prussian soldiers killed by horse kicks (Covello et al., 1982). Similar methods were used in Germany under Bismarck to enact workers' compensation statutes and retirement plans.

The concept of risk described so far largely refers to insuring oneself against possible loss, and the most accurate calculation of the costs and benefits involved. Following the same basic approach, utility theories, expected utility theories and decision making under uncertainty theories were developed beginning with Von Neumann and Morgenstern (1947), and continuing until now (Feather, 1982). The history of these approaches will be discussed in chapter three.

A totally different line of risk research was proposed by people developing personality theories. The ideas of Freud and Breuer (1895), Wundt (1874), Pavlov (1927), the drive reduction theories of Hull (1943), Spence (1956), Hebb (1955), and Eysenck (1967) as well as more recent theories, such as those of Zuckerman (1979), and Apter (1982), which focus on physiological components of risk and arousal. Yet another way of looking at risk taking is through feedback loops, homeostatic processes, and compensation behavior. There are almost as many separate ways to study and define risk, as there are researchers. I shall review risk definitions often used, and discuss their implications before presenting a risk taking definition of my own, entailing those variables of risk taking behavior elementary to the theoretical foundation of this book.

1.2 Definitions of Risk Taking

What is risk taking? Is every behavior with some degree of uncertainty risk taking? Is a person who eats a meal unaware of a carcinogenous substance a risk-taker? The person takes a high risk of being poisoned, but does not experience any risk. Is a ***conscious*** risk experience necessary for risk taking? Is the claustrophobic who goes to the movies a risk-taker? Is the millionaire who risks a penny in a game of poker a risk-taker? And what if his hand is lousy? Is the race driver who drives at 200 km/h a risk-taker, or is the beginner who drives at 25 km/h through a downtown area at rush hour a risk-taker? Is a daring chess player more of a risk-taker than a cautious motorcyclist? Is the person who does not undergo a surgical procedure more of a risk taker than the person with the same illness who does? The questions could go on and the concept would become more complex!

It is extremely hard to pin down risk taking to a simple definition. Different situations and perspectives seem to lead to different definitions of risk taking. Furthermore, different personal values of researchers seem to influence the definition of risk taking they employ. Vlek and Stallen (1980) reviewed a number of studies and listed six common definitions of risk:

1) the probability of loss,
2) the size of credible loss,
3) the expected loss,
4) the variance of the probability distribution over the probability of all possible consequences,
5) the semi-variance of the utility distribution,
6) a linear function of the expected value and the variance of the distribution of consequences.

Schäfer (1978) reviewed much of the research on risk in the area of decision making, and argued for exact, mathematically formalized theories of risk.

Schwing and Albers (1980) defined risk as a compound measure of the probability and magnitude of an adverse effect. Zuckerman (1979) referred to risk as appraised likelihood of a negative outcome. Kuriychuk (1989) split risk into Type A risk, defined as the tolerance of negative potential consequences in order to obtain higher gains, and Type B risk as arousal seeking without focus on reward parameters. Apter (1982) separated goal-oriented (telic) behavior from activity oriented (paratelic) behavior, where the rewards are intrinsic in nature. Oppe (1988) referred to risk as an expected loss, the utility of which is calculated from the combination of all input components. These components consist, however, of economic considerations only, as Oppe (1988) and Haight (1986) reduced all risky action to decision making processes which are materially evaluated.

Yates (1990; 1992) essentially treats risk taking as a decision problem. He specified the major components or the risk construct as:

1. loss,
2. the significance of loss,
3. the uncertainty associated with loss.

Yates (1992) further distinguished the losses in relation to the individual decision maker's point of reference and his/her subjective value system. The reference point takes into consideration how the person evaluates the possible consequences of a certain decision in relation to his/her current status (status quo reference). Thus, an income of $45000 may be a gain for some people and a loss for others, depending on the status quo. Further more, the same consequences (e.g. salary of $25000) may be considered a win at some stage in a person's life (e.g. as a student) and as a loss at another (e.g. after working 10 years as a general manager). On the other hand, non-status quo references leave this relation unaccounted for or refer to such general outcomes as "good health" or "happiness". Yates (1992) specified a number of such reference points for ***decision processes*** regarding risk taking, without accounting for either the attractiveness of certain risky decisions, nor the physiological and emotional benefits of risk taking behavior, nor considering the risk of missing out on a possible win (such as by not applying for a job).

Apart from differences in statistical calculations of risk, the underlying concepts of risk used by risk-researchers are also different, situation dependent, and have different shortcomings (Vlek & Stallen, 1980). For example, if risk is defined just as the probability of loss, then possible gains (winning the love of a partner after courtship) and their loss in case of no action are excluded. Considering the loss of possible gains or opportunities as risk, however, is a common phenomenon, as the excitement about winning a competition or playing a lottery demonstrates.

Even overall positive events can be considered risky, if it means loosing something else (for example: becoming a parent also means loosing independence). On the other hand, overall negative events can be considered non-risky, if the consequences are perceived as unavoidable and aiming at an overall positive goal (such as loss of mobility during pregnancy). Expectations and non-monetary gains and losses (e.g., social and physical consequences) are often not considered in definitions of risk taking (Lopes, 1987), but are nevertheless important (Oppe, 1988). Tversky (1975) even stated that his, just as most other utility theories are about attitudes to money, more than that they deal with risk in the common sense.

Even if non-material aspects are partly included in the definition of risk then the emotional experience of risk taking is still not sufficiently considered to cover all its important facets. Lopes (1987) argued that risk measured in variance, semi variance, skewness and other statistically relevant terms are neither used, nor even understood by most people and are thus hardly representative of risk-evaluations in reality. Most definitions also leave out a time factor, as they treat every moment as an independent unit, which is also highly unrealistic, as expectations about the future influence present decisions. Keller, Sarin and Weber (1986) also argued against using variance, as the variance remains unchanged if a constant of $10,000 is added to each bet of $1 or $10, while the perceived differences in riskiness become greatly reduced, since $10,001 is subjectively not perceived as being as different from $10,010, as $1 is from $10.

In addition, uncertainty in risk taking behavior is not only in the probability of an outcome occurrence ("it may happen"), but also in the perceived probability of its outcome value ("it may be useless"). Few definitions consider both aspects (Jungermann & Slovic, 1987).

Fuller (1988) postulated that risk taking is not only under the control of conscious decision making, but more often arises out of the individual's conditioning history, and this is often accompanied by the illusion of conscious involvement. Fuller (1984) argued that, especially in cases when risk taking is intrinsically rewarding, decisions will follow a sometimes not conscious compensation process. It is important, therefore, not to reduce risk taking to mere conscious decision making.

Furthermore, other terms in our language, such as danger, uncertainty and threat are often used interchangeably with risk, although some authors made sharp distinctions between them (e.g., Knight, 1921; Redlich, 1957, cited in: Schäfer, 1978; Brenner, 1983). Brenner (1983) argued that risk refers to situations with less uncertainty and equal probabilities for everyone, while the term uncertainty refers to situations when no estimate of outcomes can be made. Even in this mainly linguistic distinction, however, it becomes evident, that the term risk contains an element of uncertainty. Similar arguments hold true for other commonly used terms. As different sciences deal with different risks (e.g., biology, nuclear physics, medicine, sociology) a definition of risk taking should be applicable to all of them. Therefore, I define risk taking behavior as:

Risk taking is any consciously, or non-consciously controlled behavior with a perceived uncertainty about its outcome, and/or about its possible benefits or costs for the physical, economic or psycho-social well-being of oneself or others.

The definition refers to conscious and non-conscious behavior, outcome and consequence uncertainty, benefits and losses, intrinsic and extrinsic rewards, individual and societal risks, and the subjective experience of risk.

I shall not focus upon risk taking without a person's knowledge of it (e.g. air pollution), as the subjective experience of risk is essential for any emotional, physiological, or cognitive change in behavior. Without subjective knowledge of a risk, an individual can neither adjust to it, nor include it in any kind of cost/benefit analysis. Furthermore, the relativity of gains and losses as a consequence of subjective reference makes risk when defined as loss in any way a completely subjective concept (Yates, 1992). As the psychology of risk taking behavior refers to individuals and their perception of risk, the objective risks one has to deal with are here of minor concern.

One important aspect of risk taking will be discussed in later chapters, namely that of benefits and positive consequences of risk taking, as well as evolutionary dispositions towards risk taking. While in conscious, rational decision making processes, these aspects ***may be*** taken into consideration in average, semi-automatic, or non-deliberate processes, such aspects will be considered intuitively. After all, ***we do not just travel to arrive***.

1.3 The Multifacetedness of Risk Taking

Vlek and Stallen (1981) pointed out that different situations may require different definitions of risk. They separated large-scale societal risks, such as wars and nuclear power plants from small-scale individual risks, such as risking an accident by driving a car, and physical injuries by engaging in sports. Starr (1969), Otway and Cohen (1975), as well as Slovic, Fischhoff and Lichtenstein (1977) concluded from factor-analytic studies that perceived voluntariness, perceived control and perceived dreadfulness influence the perception and acceptance of risks.

The dimensions of "subjective"/"objective" risk (Klebelsberg, 1987; 1993) can be further differentiated by assessing risk taking on a societal or an individual level (Vlek & Stallen, 1980). Moreover, whether one engages in a task voluntarily or not, and the degree of perceived personal control over the outcome are key dimensions of risk taking behavior (Slovic et al., 1977; Fischhoff, Watson & Hope, 1984; MacGregor & Slovic, 1989).

The type of experienced consequences adds more dimensions to risk taking behavior, since physical, economic or psycho-social outcomes may have different effects on the risk taking behavior of an individual in a specific situation (Jackson, Hourany & Vidmar, 1972; Lopes, 1987).

The distinction between subjective risk (estimated and experienced by everybody) and objective risk (estimated by "experts") should theoretically provide us with differently accurate estimates. Haight (1986) pointed out, however, that experts often disagree about risk-estimates, and that the public is sometimes more accurate in its estimate of riskiness.

Furthermore, "objective" estimates are often used or even made up for political purposes (e.g., number of lives lost by drugs), so that their value for estimating risks is doubtful (Haight, 1986). Sjöberg (1980) and Fischhoff et al. (1984) argued that expert ratings and estimates of risk taking are also subjective, as is the gathering and interpretation of statistical data.

The distinction between subjective and objective risk suffers from another practical problem on the level of individual risk taking behavior. Only if an action is known or believed to bear harmful consequences, can it become subjectively experienced as a risk. Without subjective experience of risk, however, the individual cannot react to it, or consider it. If, for example, a particular mushroom dish is believed to be healthy, then its **objective** danger of being poisonous will be **subjectively** not experienced as a risk, until the body responds to the poison. If a person, however, **subjectively** believes that there is a risk involved in some action, such as using an elevator, then the **objective** riskiness of that action becomes unimportant for determining whether the person is a risk-taker.

Howarth (1988) stated that drivers' behavior correlated higher with the objective (statistical) risks of hitting a child at a particular spot, than with their exaggerated subjective risk experienced, as assessed attitudinally. This result can be explained by the finding that 99.99% of all incidences did not lead to accidents (Howarth, 1988). Thus, drivers would seem to report the risk experienced based on incidences, while their behavior would seem to correspond to the actual frequency of accidents.

Jungermann and Slovic (1987) referred to the fact that risk is an artificial construct which cannot be observed directly, therefore it lacks any "objective" reality. Thus, objective risk is a definition reached by convention, for example when referring to the number of people killed on the road in accidents per year, and then estimating the probability for this to happen in the future. An exaggerated conclusion from such a risk estimate could be that, as there are hardly any accidents with bicyclists on the German "Autobahn" this would be the safest road for bicyclists, which is of course a nonsensical conclusion. For this text, objective risk refers to either expert judgements (however wrong they may be) or to aposteriori calculated probability of outcomes.

The aspect of human's adaptive behavior to perceived riskiness is also unaccounted for in such "objective" risk estimates. Adams (1985) showed for example, that the frequency of people crossing a road significantly dropped after the road was changed such that the driven speed was much higher afterwards. Although city records showed no accidents, thus no "objective risk", their subjective risk perception led the people living along that road to certain adaptive behaviors which were considered damaging to the social life, and communication of the neighbours living on opposite sites of the road.

The separation of societal and individual risks seems to be of some importance, both for the general public as well as for the experts (Vlek & Stallen, 1980). Societal risks are typically more involuntary, uncontrollable and dreadful, such as nuclear power plant accidents, pollution, criminality, and terrorism. Individual risks, on the other hand, are perceived in just the opposite way, namely voluntary, controllable, and less dreadful, such as having a heart-attack, or filing bankruptcy. Furthermore, accepted risks are usually larger for individual than for societal risks (Starr, 1969). The general public would most likely not engage in activities that have a 10% likelihood of a fatal accident in the work place, while expeditions to the Mount Everest seem to be willing to accept these odds.

Perrow (1984) argued that many societal risks, such as DNA research, power plants, or the nuclear arms race, possess many closely coupled subsystems, meaning that if one error occurs in a sub-system, it can trigger many other errors, which may remain undetected, due to the initial error, such as a not functioning warning sign. The risks of accidents happening due to interactive failures, which can hardly be prevented, are enormous, as well as immensely complex in their causality and possible prevention.

Due to the close coupling of their individual elements, however, accidents are bound to happen (Perrow, 1984). Thus, he suggested, to leave societal risks loosely coupled (= independently controllable), to allow for independent assessment and protection of and against risks.

Individual risks, on the other hand, such as driving a car or motorcycle, mountaineering, SCUBA diving, parachuting, writing proposals, are risks which may be individually experienced as threatening, and even cost one's life, but in contrast to societal risks, the participation is normally not dangerous to others in society, or threaten the society itself, let alone the whole planet. Participation in individual risk taking activities is also typically voluntary and perceived as controllable (Slovic et al., 1978; Vlek & Stallen, 1980).

Voluntariness and perceived controllability are discussed in great detail by Fischhoff, Lichtenstein, Slovic, Derby and Keeney (1981), Fischhoff et al. (1984), Vlek and Stallen (1980), and Jeffery (1989). Vlek and Stallen (1980), as well as MacGregor and Slovic (1989) found the concept of controllability of consequences as most important for risk taking behavior and risk acceptance. As influential variables on risk acceptance the authors identified skills, appropriate equipment, opportunities for human error, recklessness, availability of rescue operations, sabotage and reversibility of effects. They also identified the factors of time, space and context of behavior itself, and its effects as important dimensions of risk taking that should be considered in studying risks.

To complicate the concept of risk taking even more, the outcomes of "risky" decisions or actions can be of multiple gains and losses, accounted for both individually and collectively. Kaplan, Szybillo and Jacoby (1974) defined risk of the prospective purchase of a certain product as the combination of: financial loss; performance loss (product does not work properly); physical loss (health, safety); psychological loss (self-image); and social loss (status).

In relation to possible effects on the individual engaging in risk taking behavior, Zuckerman (1979; 1991) referred to social and physical consequences. Jackson et al. (1972) differentiated between physical, monetary, ethical and social dimensions of risk taking. Shipley (1987) referred to the advantage of Swedish work-safety laws in including psycho-social and emotional risk factors. In this book, the two dimensions of ethical and social risk taking will be dealt with combined as psycho-social risk taking, referring to aspects of emotional experience, self-esteem, pride, etc..

Monetary risk taking will be referred to as economic risk taking and includes any material gain or loss, while physical risk taking refers to injuries, as well as positive physical experiences, such as adrenalin highs, feeling relaxed or experiencing an orgasm.

The multifacetedness of risk taking does not only refer to the behavior itself, but also to aspects of risk perception, risk assessment and risk acceptance, which will be discussed next.

Chapter 2: How Do We Perceive Risks ?

"The conditional nature of risk assessment raises the question of which standard of risk we should accept against which to calibrate human biases." (Pidgeon, 1991, p.132.)

2.1 Risk Perception and Risk Acceptance

Before risk taking behavior can occur, a risk must first be assessed, processed, and transformed to a behavioral action pattern (Fischhoff et al., 1981). The process of risk assessment underlies some obvious physical restrictions, as only perceived stimuli or cognitions can be processed, and some not necessarily obvious psychological factors.

A model describing the identification of risks as a cognitive process of identification, storage and retrieval is discussed by Yates and Stone (1992). Problems can arise at each stage of the process. The accuracy in the identification of risks has been shown to be rather unreliable, especially in complex situations or for dangers such as radiation, poison or other not easily perceptible stimuli (Perrow, 1984). Furthermore, the identification, storage and retrieval mechanisms underlie common psychological phenomena, such as primacy and recency effects, as well as familiarity habituation. That means that people familiar with a certain risk, such as driving at high speed will get used to it and accept it as a given "normal" situation and estimate the risk at a far lower value than people not familiar with the task. Yates and Stone (1992) concluded that risk takers may have different estimates and poorer judgement skills than others, as they would regularly underestimate the riskiness of their actions. The authors suggest to improve risk identification skills through processes of active search, simulation and consultation.

2.1.1 Factors Influencing Risk Perception and Risk Acceptance

In judging the riskiness of a situation or action a wide variety of verdicts is reached by different individuals, as well as the same individuals in different situations. The variety can manifest itself in terms of calibration, for example due to value induced biases which let the preferred decision appear less risky and the overconfident people choose a different anchor value (Karen, 1987).

Yates and Stone (1992) listed the following biases that have been empirically shown to influence the judgements of risk:

- Contingency judgement deficiencies {e.g. some people believe in causal relationships of astrological patterns and success}.
- Value biases {e.g. people overestimate their likelihood of success if they desire the outcome, such as winning $1.000.000 in a lottery}.
- Personal role biases {e.g. Illusion of Control, such as guardian angel for motorcyclists (Rheinberg, Dirksen & Nagels; 1986) leads people to underestimate the riskiness of actions}.
- Perceptually based biases {e.g. large objects (locomotives) are perceived as slower than small ones}.
- Mood and individual difference effects {e.g. Johnson & Tversky (1983), showed that people in a depressive mood estimate the probability of failure to be higher}.
- Level effects {Fischhoff and MacGregor (1991) showed that the wording of the question led to different risk estimates}.

Further differences in judgemental skills become obvious with the degree of knowledge and expertise of the risk assessor. Lichtenstein, Slovic, Fischhoff and Combs (1978) reported that the correlations between lay-peoples' individual assessments of riskiness and statistical frequencies of fatalities lies only at r=.50 to r=-.62. They concluded that either lay-people commit serious errors in their assessment, or they include other aspects of risk in their assessment than experts, who base their judgement mainly on fatality statistics (Jungermann & Slovic, 1987). Von Winterfeldt, John and Borcherding (1981) reported, however, that subjects were relatively correct in estimating fatalities directly, rather than through estimating risks. Slovic, Fischhoff and Lichtenstein (1980) reported that peoples' fatality estimates did not differ significantly from experts' estimates, but that the correlations between actual accident frequency and perceived risk were only r=.26-.60, thus explaining 6%-36% of the variance. They concluded that non-experts assess risks qualitatively differently, by including the dimensions of controllability, voluntariness, dreadfulness and whether a risk is known or not. Hansson (1989) also argued against the sole reliance on uni-dimensional mathematical assessment of risks based on expert's judgement of material costs involved. Weber (1988) gave a comprehensive review of such single-dimension situations.

Factor analytic studies (Fischhoff et al., 1981) have supported these factors, finding three dimensions of risk-assessment and acceptance, namely "dread risk", "unknown risk" and "exposure" (e.g. how many people are in danger). Vlek and Stallen (1981) found two factors, generally corresponding with "dreadfulness" and "unknown risk".

Risk assessment also depends on the situation and context of the risk that is to be judged, as well as the situation and context of the judges themselves (Von Winterfeldt & Edwards, 1984). Risk situations concerning health and food were evaluated differently from risks in industrial contexts and relatively unpredictable risks, such as nuclear power, or genetic research. Cultural and psychological contexts are further factors that influence risk assessment (Otway & Von Winterfeldt, 1982).

The factors of voluntariness and controllability also influence the judgement of risk. Supporting the difference between societal and individual risks, the importance of voluntariness and controllability could be found in a study by Von Winterfeldt, John and Borcherding (1981). These authors reported lower perceived riskiness for motorcycling, stuntmen work and auto-racing, than for nuclear power and air-traffic accidents.

Renn (1981) reported a study on voluntariness and perceived negative effects. One group of subjects was allowed to choose between three types of pills, while the other group was administered these pills. Although all pills were identical, the voluntary group reported significantly fewer "side-effects", than the administered group. The importance of the voluntariness of an action and its evaluation are also apparent in our law systems, where an accidental, or involuntary killing is judged differently from a planned murder (Jungermann & Slovic, 1987), even if more people die as a result of the accident. Differences in the consequences of perceived risk were found by Slovic, MacGregor and Kraus (1987), who reported that subjects' estimates of the riskiness of a car-defect influenced their decision to buy another car of that brand.

The influence of "emotional frames" (affective context with induced emotions) on risk assessment and acceptance could be shown by Johnson and Tversky (1983). In the frames used by these authors positive and negative emotions were induced through descriptions of events such as either the personal success, or the death of a young man. They found that subjects with induced negative feelings judged the risks of accidental and violent fatality rates as significantly higher, regardless of context, than subjects of the positive emotional group.

Similar influences of situational frames on risk-evaluations have been reported by Kahneman and Tversky (1979; 1984), and McNeil, Pauker, Sox and

Tversky (1982). The latter authors reported that if they phrased the outcome of a therapy as 68% probability of survival, 44% of the subjects chose it, as opposed to only 18%, who chose the same surgical or radiation therapy, if the outcome was phrased as 32% probability of death, which are both mathematically equivalent.

Results like the ones cited above further complicate the process of risk acceptance. Often subjects choose an anchor value (Lopes & Ekberg, 1980) to judge the acceptability of risks, especially when dealing with cumulative risks over time (Svenson, 1984). Fortenberry and Smith (1981) reported that people chose the same risk of financial or electric-shock punishment for others, as they chose for themselves.

2.1.2 Research on Risk Perception and Risk Acceptance

Fischhoff et al. (1978) listed emotional, cognitive, linguistic and statistical factors influencing risk acceptance, and concluded that large scale societal risks should be accepted only when a well informed public makes the decision. This raises the political problem of whom to inform how about what. Furthermore, as news-media report disastrous events very often and very dramatically, these events become over-represented in the public's eyes and memory and the risk assessment, as well as the risk acceptance are influenced significantly (e.g., nuclear power plants) (Fischhoff, 1977).

Christensen-Szalanski, Beck, Christensen-Szalanski and Koepsell (1983) reported that although the "experts", namely physicians, were more accurate in their estimate than undergraduate students, both groups systematically overestimated risks associated with several diseases. The authors attributed this finding to an increased exposure to these diseases, and to a greater medical journal coverage of the medical students.

Short (1984) argued that risk acceptance largely depends on the trust in the institutions presenting the "expert" rating, and by the previous failures and success of those institutions. Slovic et al. (1980) and Jeffery (1989) suggested to improve methods to understand, assess and communicate risks.

Kasperson (1986) discussed necessary aspects for including the general public into a risk taking decision process on a societal level. Fischhoff et al. (1978) suggested to use a technique of paired comparisons of risk taking activities and base societal risk-acceptabilities on that.

Bajgier and Moskowitz (1982) argued for the use of an interactive model of attitudes, beliefs and risk/benefit formation. These factors are, however, also influenced by perceived controllability and dread, as well as knowledge of risk. Further methods to improve knowledge about possible risks and improving decision making skills are discussed in Yates and Stone (1992).

Fischhoff's (1977) argument for a rational decision making process by a cost/benefit assessment, seems highly valid, but not very representative of how risk taking behavior and risk-decisions exist in reality. Slovic et al. (1977), Fishburn (1988) and Fischhoff et al. (1982) stated, that often decisions are made irrationally, from a benefit maximization point of view.

Zimolong (1985) reported that greater exposure to areas of high falling-injury risks reduces the estimate of its occurrence significantly below its actual occurrence. Gerking, De Haan and Schulze (1988) found that subjects were significantly more willing to accept higher wages for more dangerous jobs, than to pay for making the job safer, thus maximizing economic benefit in favour of physical costs.

Fischhoff, Furby and Gregory (1987), Lopes (1981) and Vlek and Stallen (1980) suggested to include more personal and intrinsic reward aspects in cost/benefit analyses to understand the processes of risk assessment and risk acceptance more completely.

Machlis and Rosa (1990) discussed the concept of desired risk in contrast to tolerated or dreaded risk. They concluded that in many situations increased risks function as an incentive, rather than as a deterrence. They stated that amusement park operators have reported the ride becoming more popular when reopened after a fatality. When a Norwegian ferry sank and the passengers were set afloat on icebergs for 36 hours, the company experienced the greatest demand on their trips ever.

Machlis and Rosa (1990) concluded that the concept of desired risk changes the perception and acceptance of risks and demands different conceptual models to explain risk taking behavior. Jermier et al. (1989) supported the assumption of desired risk by reporting that for police officers on patrol the physical danger of being attacked or killed was perceived as job enrichment, while for police officers engaged in administrative duties, this risk was perceived as dreadful.

Jungermann and Slovic (1987) stated that individual differences in perception, evaluation, and acceptance of risk can be found for students, technicians, and environmental activists. Aschenbrenner and Wurm (1985) showed that SCUBA-divers perceive risk differently from non-divers, and stated that differences in risk-assessment, risk-acceptance and risk taking are most likely influenced by personality factors.

Vlek and Stallen (1980) argued that physiological, behavioral and cognitive measurements of perceived riskiness often differ, due to cultural, situational and individual differences.

2.2 Cross-Cultural Aspects of Risk Perception

After having discussed the importance of risk perception, risk estimation and risk acceptance the question is raised as to what individual, societal or cultural factors influence this perception and acceptance of risks. Cultural theorists have proposed that individuals choose what to fear and how much to fear it in order to support their way of life. Such perceptions are influenced by cultural biases and social relations (Dake, 1991; Douglas & Wildavsky, 1982).

2.2.1 Research Relating Culture and Risk Perception

Following this idea one first has to define what is meant by the term culture. Pidgeon (1991) referred to three different definitions, all of which have received acceptance and have been used in cross-cultural research:

1. culture is determined through regularly occurring behavior within a technological, economic, religious, political, familial or other institutional domain within a population,
2. or culture as a system of meanings and symbols, of thoughts, rules and ideas in the heads of members of a population,
3. or a combined definition of culture as the collection of beleiefs, norms, attitudes, roles, and practices shared within a given social grouping or population.

The influence of culture (e.g. compared countries include: USA, Germany, Hungary, France, Norway, New Zealand, Australia, Sweden, Holland, Japan, Russia, China, Hong Kong, Britain, etc.) on individual risk perception and risk acceptance has been extensively documented by Fischhoff et al. (1978), Rohrmann (in press), or Slovic et al. (1980). Usually, in such studies a sample of possible hazards or risks were presented and rated according to a set of risk aspects. Sometimes a single risk source was presented, such as nuclear power plants or toxic waste (Vari, Kemp & Mumpower, 1991).

Additional cultural aspects examined were social, political, religious or economic factors (Vari et al, 1991), or worldviews, such as individualism, versus egalitarianism versus belief in hierarchies (Dake, 1991). These last personality-trait-like aspects as well as the influence of prior experiences, beliefs, judgements, norms and level of exposure have led to significant behavioral and perceptual differences both between countries as well as within countries, comparing ethnic or socio-economically different populations (Vaughan & Nordenstam, 1991).

Kleinhesselink and Rosa (1991) found that Japanese perceive atomic power as a dreadful, but not as an unknown risk, while for Americans atomic power is a predominantly unknown source of risk. The authors attribute these differences to different exposure, such as to the atomic bombs on Hiroshima and Nagasaki in 1945. However, similar differences were reported between Hispanic Americans and white, egalitarian American residents of the San Francisco area. Thus, individual differences and personality aspects may play an equally high role in risk perception as general cultural biases do (Rohrmann, in press).

McDaniels and Gregory (1991) suggested to structure risk perception and decision making into an individual, an organizational and a governmental component and examine the influence "culture" has on these aspects. McDaniels and Gregory (1991) further divided cultural risk comparisons between individuals, organizations and governments into:

- decision making practices,
- financial risk behavior and,
- physical risk behavior.

They suggested to examine a number of research issues on the individual level (e.g. Does culture influence the use of probability, entrepreneural patterns, protective behavior?), answering questions as to whether different cultures influence risk perceptions. Questions on the organizational level, for example, refer to organizational objectives, incentives or responsibilities, while questions on the governmental level are aimed at determining the influence of rules and regulations, fiscal policies or interest groups on risk perceptions, and many more. The authors offered some different risk strategies and risk preferences on the organizational level, such as that Chinese executives are on average more inclined than Canadians to adopt a face-saving option, or a long-term business arrangement. Further differences on the individual level were also specified.

These and similar findings led Johnson (1991) to formulate cautious warnings about the causal attribution of cultural differences to risk perception and risk acceptance. He worries about the widely spread differences in the definition of culture which makes it almost an all encompassing label. Moreover, considering the differences in opinions and behaviors of subpopulations or individual business organizations within a country adds further problems to a clear cut measurement of culture or its effects on risk perception and risk acceptance. Also, the samples studied are usually small and not representative of the cultures as a whole and often causes and effects are not seperated properly (Rohrmann, in press).

Yet, one of the most consistent findings across cultures and within cultures, is that usually a two factor structure of risk perception evolves, namely the concepts of dreadfulness and unknown risks discussed in chapter 2.1. Also, the question of voluntariness and controllability are recurring topics in virtually every study on risk perception (Kasperson, 1986; Kleinhesselink & Rosa, 1991; Rohrmann, in press; Slovic et al., 1980). Although these factors often emerge, they receive different priorities in the countries examined.

Whether these priorities are culture dependent or situation dependent remains a question of debate (Johnson, 1991). For example, the finding that in estimating the hazards of toxic and radioactive waste disposal, British people focus more on transportation risks, Hungarians more on operating risks and Americans more on environmental risks. These differences are attributed to cultural differences (Vari et al., 1991) but may just as well be the result of a perceived population density in Britain, little operating reliability in Hungary and the Three-Mile Island environmental threat in the USA, which are situational factors.

In other words, although the major components of risk perception seem to be cross-culturally consistent, but their degree, direction and interpretation is very much dependent upon individual, subgroup and societal differences. Further resaerch has to establish more of causal examinations and explanations to determine the contribution of national culture to the concepts of risk perception and risk taking behavior.

2.2.2 Safety Culture

Pidgeon (1991) referred to the organization as a cultural subgroup with its own set of norms, rules and regulations. The notion of *corporate culture* has received wide attention in the last decade (Peters & Waterman, 1983). In relation to risk perception and risk taking the notion of a ***safety culture*** has become an important issue in research and in applied safety work.

Pidgeon (1991, p.135) defined safety culture as a constructed system of meanings through which a given people or group understands the hazards of the world. This system specifies what is important and legitimate, and explains relationships to matters of life and death, work and danger. A culture is created and recreated as members of it repeatedly behave in ways that seem to them to be natural, obvious, and unquestionable ways of acting, and as such will construct a particular version of risk, danger, and safety. Such versions of the perils of the world also will embody explanatory schemata to describe the causation of accidents. Within an organization the tacit and explicit rules and norms governing safety are at the heart of a safety culture. Major components are thus:

- rules for handling hazards,
- attitudes toward safety, and
- reflexivity on safety practice.

Organizations with an elaborate safety culture do emphasize the importance of common visions, goals, standards and behaviors in risk taking behavior. As risks are unavoidable within the context of work, an optimal balance of risk taking and control of hazards has to stricken. Vlek and Cvetkovitch (1989, pp.302-303) stated:

"adequate risk management is a matter of organizing and maintaining a sufficient degree of (dynamic) control over a technological activity, rather than continually, or just once, measuring accident probabilities and distributing the message that these are, and will be, "negligibly low". Thus more often than not, "acceptable risk" means "sufficient control".

Accurate communication of risk is thus the necessary tool to develop adequate risk perception, risk estimation and risk taking behavior (Rohrmann, 1992). Risk communication serves simultaneously to inform people about expert estimates. It helps people who perceive themselves threatened to find a forum for discussing their worries. Risk communication between subgroups can reduce cultural differences in risk perceptions and influence governmental policies regarding for example environmental dangers.

One result of the research on risk perception is that the individual interpretation, personal preferences and previous experiences are major contributing factors. I shall thus now focus on personality aspects of risk taking.

Chapter 3: Who Engages In Risk Taking Behavior ?

In Chapter 1, it was mentioned in some theories that risk taking behavior may be influenced by individual differences. Traditionally, men have taken more physical risks than women (e.g., wars), and especially young men are known to show "mating and other hierarchy-establishing behaviors" that includs high-risk actions, such as driving a motorcycle on the hind-wheel, or engaging in duels.

World history, however, shows that large individual differences in type and degree of risk taking exist among well-known risk-takers (e.g. Socrates, Caesar, Napoleon, Darwin, Gandhi, Mother Theresa). What determines the preference for a social risk over a physical risk? Was Gandhi a "born" pacifist or did he choose passive resistance as the solution of a cost/ benefit analysis? Are "heroes" genetically different from "cowards" or did different life-experiences determine their actions?

In the following, I shall summarize arguments in favour of a genetic component of risk taking behavior. It is important, however, to realize that genetic disposition does not imply either invariability or a specific type of behavior. To illustrate, although the desire to eat is certainly genetically controlled, when, what, how much and how often we eat is very much flexible and under the individual's control.

When applied to risk taking, my argument that risk taking is a genetically determined action underlies the same logic as eating does. Who takes what risks, why, when, and for how long is flexible, but every living person has to take risks in one form or another, and thus, every person chooses to take risks to optimize one's chances for survival and personal welfare.

3.1 Risk Taking and the Theory of Evolution

If risk taking behavior has an invariant, inherited personality-trait component, it would have to have a genetic basis. Buss (1991, p.461) argued:

"All observable behavior is the product of mechanisms residing within the organism, combined with environmental and organismic inputs that activate those mechanisms."

Genetically predisposed mechanisms and behaviors will thus play an active role in every action. The degree of influence, however, is situationally and individually different. Genetically evolved mechanisms usually evolve with an increased benefit for survival. A genetic risk taking factor would thus have to fulfill a beneficial function in evolution, and lead to a selection process favouring organisms possessing a genetic pattern for risk taking. The following chapter will provide the evolutionary background for the assumption that risk taking behavior is inevitable, essential to evolution and, therefore, at least in part genetically determined, while different ways of expressing risk taking behavior are environmentally determined. Modern approaches to "psychobiology" take both aspects into account, as Dewsbury (1991) showed in a historical review of psychobiological concepts.

The theory of evolution (Darwin, 1859/1968) is based on the assumption that if the resources necessary for survival are scarce, organisms compete with each other, both between-species, as well as within-species. The best equipped organism will gain the resources, and thus increase its chances for successfully producing and raising offspring, which in turn will proliferate the genetic pattern that led to the advantage in the first place. This type of organism will thus have a higher likelihood of survival. Dawkins (1976; 1986) described an extension to Darwin's theory by identifying genetic survival, rather than survival of the species as the motivator of survival behavior. According to Dawkins (1976), it is not the survival of the species that is the primary goal of an organism, but rather the survival of genes that is automatically coded into every organism. The body is the phenotypic expression and carrier of the genes, and as such it is responsible for both the gene's, as well as for its own survival.

Members of the same species compete with each other for scarce resources and for the best reproductive partner, as their individual gene-set programs them to strive for optimal survival conditions (Dawkins, 1976; Zuckerman, 1990). Darwin (1871) differentiated between strategies that give organisms survival advantages and those giving them reproductive advantages. Although connected through the survival of genes, the two areas have different behavioral consequences. Mating is a universally observed behaviour and more than 90% of all people in all societies get married at some point in their life (Buss & Schmitt, 1993). It thus takes a very important role in our lives. Most people will remember feelings of uncertainty and arousal when establishing a relationship with a prospective partner. In a time where sexually transmitted diseases (e.g. AIDS) are an increasing problem, mate selection has to be considered as one form of social and physical risk taking.

3.1.1 Evolutionary Aspects of Human Mating as Risk Taking

In the competition for obtaining the desired sexual partner humans and otheranimals alike engage in very dangerous behaviors. Especially males engage in fierce fights on the physical level or in modern cultures on verbal and financial levels to win the favour of the respective partner and beat their opponents. Incredible risk to one`s health, prestige, and virtually all important and valuable areas of our lives are taken when one ***"gets blind with love"***. This type of risk taking behavior has been reported from the very beginning of human history. It is at the center of many novels, biographies, fairytales and even in modern stories, be it books or films.

Furthermore, as many personal experiences also seem to validate the importance and consistency of this form of risk taking, I want to move to the evolutionary roots of this incredibly *potent* motive.

Due to the nature of genetic and organism reproduction, it is only when organisms have very similar genetic patterns, namely within the same species, that they can produce offspring. The more offspring an organism produces, the higher the likelihood of the survival of its genetic pattern. Consequently, the more similar the genetic patterns of organisms are, the less likely is an organism to destroy its carrier. The killing of direct family members is, therefore, even more rare in nature, than killing members of the same species. Child murder should have the lowest frequency, because parents have "invested" their genes and energy into them. Wilson and Daly (1985), and Daly and Wilson (1988) showed that this assumption is supported by USA homicide rates.

Howard (1986) reported that people with more siblings, less parental bonding, who are younger, and male, commit more crimes, and more serious crimes. This finding was also supported by Wilson and Daly (1985), who argued that psychopathic behavior is being developed through an interaction of a genetic predisposition to social withdrawal and early exposure to an uncontrollable environment, thus pointing out the importance of the individual's perceived control over the environment. Cadoret, Crain and Crowe (1983) also referred to genetic influences that may support the development of antisocial behavior.

We thus have environmental, developmental and possibly genetic factors influencing forms of social or anti-social risk taking behavior. The amount of energy, effort and personal sacrifices invested into getting and caring for offspring seems to influence the attitude and care-taking behavior of parents. Studies examining the mechanisms involved commonly refer to it as the ***Parental Investment Model*** (Trivers, 1985; Feingold, 1993). Parental investment is defined (Trivers, 1972 p. 139) as:

"Any investment by the parent in an individual offspring that increases the offspring's chances of surviving (and hence reproducing) at the cost of the parent's ability to invest in other offspring."

This model implies that among mammals, males and females of the same species have to employ different strategies to ensure that they select their optimal reproduction partner. In most species, the females are responsible for pregnancy and raising of the new-born (Feingold, 1993). In cases where this role is fulfilled by males, mating strategies are reversed (Dawkins, 1976). Thus, for males the best strategy to increase the probability of survival of their genes is to impregnate as many females as possible. Females, on the other hand, have to choose a different strategy in mate selection, as they are investing much more time, resources and energy during pregnancy and the initial raising of the offspring (Buss, 1985).

Therefore, for a female it is much more important to select an excellent reproductive partner, who possesses a genetic set that maximizes her own chances of survival as well as that of her offspring, and that minimizes her energy expenditure. That is, the ideal partner would possess such characteristics as physical strength for protection against aggressors and illness, or a particular pattern of colors that is advantageous, and a willingness to share expenditures of the pregnancy and child raising, such as providing food. To obtain a mate, males have to prove their worth in these and many related aspects, such as competing with other males desiring the same reproduction partner (Dawkins, 1976).

Buss (1988) argued that based on Darwin's (1872/1934) theory of sexual selection, males should be young and healthy and compete in combats to prove that they are strong and capable to defend the female and their offspring. Buss (1988) showed, that male students put more weight on displaying physical abilities and properties, thereby indicating their worthiness as mates, and on competing successfully with members of the same sex. In another part of the study, Buss (1988) showed the same effect for married couples, who assessed their own and their partners' behavior retrospectively. The optimal partner for a female is therefore, according to Buss (1988), a male that can provide material to provide protection, food, shelter, and a physically strong and healthy body (Buss and Barnes, 1986).

In a review of studies, Buss and Schmidt (1993) showed that men are evaluated more highly than women on characteristics such as:

ambition, industriousness, social status, commitment to other women, age, athleticism, intelligence, material possessions, territory, ability to protect, and ornamentation.

Furthermore, *good financial prospect* seems to be a cross-culturally important feature for men to fulfill in their modern role as care-taker and provider. Kenrick, Groth, Trost and Sadallah (1993) further discussed these features as differentially important for varying purposes of mate selection. When selecting a long-term mate these features are very important for women while for a short term or mainly sexual mate the factor attractiveness seems to be of greater interest for both sexes. The notion of attractiveness, however, possesses different associative links for men than for women. While both sexes consider physical health as attractive, women find "mature and dominant features" attractive in men, while men consider "youthfulness and physical expressions of fertility" as attractive. (Kenrick et al., 1993). Accordingly, "dominance, powerfulness and aggressiveness" are considered as increasing mens' attractiveness (Wiederman & Allgeier, 1992) while that is not so for the attractiveness ratings of women in the eyes of men.

For the male on the other hand, the female has to be mainly healthy and physically optimal to provide a good gene-pool, as well as to survive the pregnancy and raise his offspring well. Furthermore, the female has to show off her fertility and sexual ability, to make male "investments" of energy and resources worthwhile.

The risk of mating with an inappropriate partner has less costly consequences for males than for females. Males normally do not become pregnant, can leave immediately after an impregnation, which costs very little, and is intrinsically rewarded by an orgasm. Yet, as females force males to invest energy and resources in being selected, the male investment becomes bigger, and thus the males are less likely to leave. Buss (1988) showed that female students and married women alike put more emphasis on appearing attractive, healthy, sexually capable and available, and on emotional caring, kindness and understanding. He also found both sexes to perceive mating strategies as differentially successful for men and women.

Men are more inclined to have short-term, low involvement sexual relationships and place significantly lower standards on their sexual mating partners. An evolutionary optimal strategy would be to "produce" as many children (gene-carriers) as possible without the investment of rearing them. Thus, attractiveness and fertility as well as easy accessibility are the main selection factors for men in short term relationships (Kenrick et al., 1993). Women, on the other hand, already place in short-term or purely sexual relationships high standards on character and abilities to provide for possible offspring. Thus, according to Feingold (1993) even not so physically attractive but wealthy men have an increased chance for mating.

"The unattractive woman, like the poor man, is kept off the playing field" (Feingold, 1993, p.134).

Also, Buss and Barnes (1986) as well as Kenrick et al. (1993) showed that males put increasingly more weight on womens` attributes of physical attractiveness, being a good cook, showing child-rearing interests and frugality with increasing interest in long-term or marital relationships. The longer the planned investment the closer become the standards the mate has to fulfill. In these studies, women preferred men who are: honest, kind, dependable, considerate, fond of children, tall, and have good earning capacity, thus reducing the risk of being left alone with child-rearing responsibilities.

Such preferences confirm the suggestions based on evolutionary theories, but are of course also in line with current and past social values, and sex roles in our societies, therefore supporting a learning theory approach just as strongly. The findings, however, tend to confirm the traditional roles of males and females, at least valid in present time.

Feingold (1993) further presented a meta-analysis of 34 studies testing the parental investment model and finds convincing support. He does, however, point out a number of problems that need to be addressed in future research. Firstly, he differentiates between

- romantic attraction (affective dimension)
- self-perceived mate-preferences (social-cognitive dimension), and
- dating/mating choices (a behavioral dimension).

These dimensions lead to different consequences and are secondly related to factors such as:

- social desirability (choosing the perfect mate),
- social approval of significant others (approval or even envy of friends and family),
- accessibility of possible partners (attractive partners have a larger selection pool), and
- the interactions of several of such factors.

Feingold gave further reasons why the gender differences between romantic attraction and romantic behavior may not be identical. He described the imbalance in *gender-related courtship roles*. Men traditionally shoulder the costs and risks of dating, both socially (risk of rejection when asking the woman out) and financially (traditionally men pay for meals and other expenses). These investments lead to different selection criteria for asking women out (attractiveness) and accepting it (financial prospect).

Cultural roles differ. Traditionally women worked in the house, while men were responsible for earning the financial basis for a family. Thus, the socio -economic status of men seems more important than that of women. Many modern women have jobs and work after being married, yet, the data show that women are more willing to sacrifice their careers, to move with their husbands and to take longer periods of maternity leave than is true for men.

Cultural mating systems, such as arranged marriages, shift the risks in mate selection from the wishes of each individual partner to the assessment capabilities of the parents. The relationship between attraction and behavior is thus not to be expected to be consistent.

Furthermore, a majority of people mate with partners from the same *socio-economic status*. Due to reasons of proximity and opportunity (same schools, clubs, streets, bars, jobs) as well as social approval of friends and family the romantic attraction and behavior may differ.

Finally, *love styles* seem to be different with men and women. While men typically prefer to move faster towards a sexual relationship, women prefer to have a longer evaluation and familiarization time. Thus, the behavior varies from pushing to resisting (Feingold, 1993). Again, recent social developments make it much more socially acceptable for women to reverse roles so that traditionally "male" behavior patterns is found in women, too.

The complexity of these issues has not been studied thoroughly yet. Furthermore, the possibility of an Halo effect (changing the judgement of one feature due to observing another characteristic of that person) is very difficult to assess and as most questionnaires, or interviews claim independence of measures they can not even pick up on such issues if they occur.

Issues of changing societal roles, pragmatism in mate selection, and childhood developmental issues are hardly controlled for or addressed in modern research on mate selection. As, however, these factors will play a role in the risk perception and risk appraisal during the initiation of romantic relations, following-up on a first contact, showing long-term commitment and in ending relationships, they need to be better understood.

To add to the complicated picture, Lykken and Tellegen (1993) examined the actual mate selection factors of modern monozygotic twins and compared them to dizygotic twins and not related couples. They set the studies up to test assumptions based on different models.

Idiographic models, such as the one proposed by Buss (1988) suggest that there are certain evolutionary derived, rational and thus "lawful" choices why people select their mates. Thus, identical twins should show similar choices in mates and should be rated as similar in attractiveness by their co-twins' mates. What they found, however, on a sample of more than 100 twins from the Minnesota Twin Study was that twins rate their co-twins taste in clothing, vacation jobs and houses as very similar but not in their choice of mates. The mates of twins do make very similar ratings of their partner's co-twin, rating them very similar except in their choices of partners. The partners of dyzigotic twins consider their partners co-twin even as more negative than positive. These data point very much against a stringent and lawful criterion to be at work in human mate selection.

Lykken and Tellegen did, however, argue that with the rather unspecific criteria derived from evolutionary considerations, such as healthiness, attractiveness etc., 75% to 94% of the women of each subgroup could be rated as possible mating partners. Thus, the final factors of choice have to lie somewhere else. They then refer to animal models where pair-bonding is dominant, such as in geese, wolves or gibbons. They assumed that at early stages of bonding an imprinting-like process occurs which is partly triggered by potential partners positive responses to one's advancements.

Liebowitz (1983) associated these romantic infatuations with increased specific neurotransmitter activities creating sensations of euphoria and optimism. This process is estimated to last for about 2-3 years, after which statistically most relationships split. On the other hand this extended period of attraction provides the necessary time to establish so called "compa-nionate love" which relies on mutual experiences, similarities in interests and tests of reliability and fit. According to Lykken and Tellegen (1993) choice of mates can not be explained or even predicted by rational lawful models but are at present still beyond the grasp of scientific minds. This uncertainty and unpredictability, however, further increases the risks and the benefits associated with mate selection.

Thus, at present mate selection can be considered as risk taking behavior, varying in form and degree between sexes as well as within sexes. Evolutio-nary beneficial concepts coincide with cultural norms and socially expected role models. As behavioral predictions from either theoretical approach often coincide in their explanations and cazsal attributions, it is at present hardly possible to differentiate clearly between them. Kenrick et al. (1993) suggest therefore, to link evolutionary and social theories to gain a deeper understanding into *why* (evolutionary dispositions) and *how* (social and cultural expression) mate selection takes place.

3.1.2 Risk Taking Behavior and Optimal Chances of Survival

Competition for resources requires a certain amount of skill, or mastery, an ability to find these resources and the desire and curiosity to look for them (Buss, 1988). To display such curiosity, however, is not without its dangers. Changes in the environment are unavoidable. Examples are: seasonal changes, weather changes, influences of other predators and competitors, changes in the availability of resources, and changes such as the birth and death of family members. While encountering such changes, every organism is constantly exposed to a changing level of uncertainty, danger, or risk.

Threats and dangers require from the organism to engage in a type of fight-flight behavior, (Buss, 1988), which in turn requires the body to function optimally. To be sleepy and unprepared increases the chances of being killed. Therefore organisms reacting with heightened arousal, attention and preparation in response to a possible threat increase their likelihood of survival (Easterbrook, 1959; Hamilton, 1979).

Apart from increased arousal in unknown situations, the likelihood of survival is also increased if an organism possesses the ability and desire to influence and control the environment. An unknown environment is an uncertain one with possible threats to survival. Strategies that are generally employable to control unknown situations, while simultaneously being intrinsically rewarded and genetically programmed to be used in such situations, would prove advantageous. On the other hand, the great variety of possible threats makes the ability of flexible adaptation to the environment just as important for survival (Crawford and Anderson, 1989).

In addition to the desire to influence and control the environment, however, the willingness to take some risks in order to acquire necessary resources is also advantageous, as most often scarce resources can only be competed for under the risk of injury or death. Any organism genetically programmed to take an ***optimal amount of risk***, has, therefore, a great advantage and is more likely to survive and to have more offspring. This in turn increases the frequency of its particular genetic code, and its type of carrier body.

Optimal risk taking, however, is determined by the environmental situation and the individual's abilities (Buss, 1988). An individual should be capable to assess relatively accurately the danger in acts aimed at gaining access to resources. In addition, practicing these situations is necessary to learn the essential skills. Individuals with the genetic program to learn the skills in a safe situation, like in playing or when competing with peers, should have a higher likelihood of survival, as they are better prepared when a dangerous "real" situation occurs.

Physiological Reward Systems for Optimal Risk Taking

An organism which provides an intrinsic reward system for taking risks up to an optimal level and a reward system for punishing risk taking beyond optimal levels, has an even higher likelihood of survival, as if either risk taking or control alone would be rewarded. Olds and Fobes (1981) supported the assumption of neurologic, homeostatic, intrinsic self-reward mechanisms by studies on self-stimulation. They located reward structures predominantly in the hypothalamus area. The hypothalamic, and other reward structures in the brain were shown to be connected to activities concerning food intake, sexuality, and temperature regulation (Olds & Fobes, 1981).

Changes in arousal and physiological rewards for optimal risk taking behavior lead to emotional experiences. Also, as soon as a neurological compensation process is initiated to reach the subjectively optimal level of risk, this process itself should lead to a positive feeling. There is some evidence relating evolution to emotion and risk taking by suggesting that emotions are genetically determined motivating action patterns influencing risk taking behavior (Toda, 1980; Frijda, 1988). From this it can be concluded, that risk taking is physiologically, emotionally and cognitively controlled, experienced and rewarded.

Consistently taking too high risks, even if subjectively optimal and intrinsically rewarded, will eventually result in extrinsic punishment in form of death or serious injury, while taking objectively too few risks will eventually result in not getting the necessary resources. Faulty risk assessments can lead to extinction, either way.

It is also logical to postulate that organisms, such as human beings, who grew up under different conditions have developed different survival strategies, and different optimal levels of risk taking. Prehistoric humans living in environments rich in vegetation and with little dangers from predators were hardly encouraged or forced to expose themselves to any of the dangers around. They could afford to wait until the threat was removed before continuing to look for food. This type of environment would have encouraged patient and risk avoiding temperaments (Crawford, 1989).

On the other hand, an environment with scarce resources, and more predators would have forced its inhabitants into a fierce competition for food. If the same type of food is searched for by humans and their predators, conflicts are likely to occur. To provide themselves and their offspring with essential resources, like food, these humans had to take considerably higher risks, both to outwit other predators, as well as to acquire the necessary resources, because their prey would also be more cautious and harder to overcome. This type of environment would have encouraged more risky behavior, as a patient, risk avoiding person would have been less likely to gain access to the necessary resources.

Organisms with a less aggressive and less risk-encouraging genetic pattern would have been more likely to survive and reproduce in the first scenario than in the second one. The offspring of both groups, would then start from a different baseline in their risk taking tendencies and, depending on their own new environment, would pass on their genes if they survived. Crawford (1989) supported this assumption and argued for different strategies of raising offspring, with a less caring strategy in a rich environment, and a caring one in an environment scarce in resources. Large genetic and behavioral individual differences are, therefore, expected to appear as a consequence of different surroundings.

Because environmental conditions can change drastically over space and time the variety of risk taking tendencies within a population over thousands of years will logically be rather big, as well as very mixed, due to migration and inter-tribal mating (Dawkins, 1986). The variety in risk taking tendencies is further increased by the development of better protective equipment and strategies. These allowed for less risky hunting or searches for food, better protection of the ill, injured, or old members of a group, who in return pass on their knowledge of risks and dangers. Different genetic sets can prove advantageous to their owners, depending on their specific situation (Crawford & Anderson, 1989).

Blanchard and Blanchard (1988), for example, described risk assessment situations that typically involved the attacker disguising himself to attack successfully and the defender focussing attention to any change, to avoid being surprised by an attacker. They argued that both patterns are highly motivated and form an optimal fit between risk assessment and risk avoidance. Unidentified and uncontrollable dangers often result in a freezing reaction, combined with high anxiety, which Blanchard and Blanchard (1987) distinguish from fear of a specific threat. The same freezing phenomenon was basis for the "learned helplessness" model by Seligman (1975).

Blanchard and Blanchard (1988) pointed out that an aggressive offense is very situation specific, while defensive behavior is similar in all vertebrates. Flight, freezing, defensive threat and attack are the usual defensive responses which are also followed in this order, if a predator is perceived. Blanchard and Blanchard (1987) showed, however, that in domesticated animals this pattern has been disturbed by selective breeding.

To ensure optimal behavior in risk assessment, defense, offense and their interactions, Blanchard and Blanchard (1984) argued that lower mammals and human beings undertake a cost/benefit analysis on an emotional basis.

The concepts of fear, aggression and other related emotions are derived from their behavioral and physiological correlates. Blanchard and Blanchard (1988) offered a comprehensive review of the neural structures involved in defensive and offensive action, as well as in the identification of those complex environmental stimuli necessary for accurate risk assessment. They identified the amygdala, hypothalamus and the midbrain as responsible for elicitation, modulation and organization of such behaviors. The authors further reported that stimulating the human amygdala typically elicits reports of fear or anxiety, with little or no anger.

Blanchard and Blanchard (1988) cited studies favouring testosterone, and adreno-corticoids as the hormones responsible for offense, defense and related emotions, while Panksepp (1986) reviewed studies on the pharmacological control of emotions. Risk taking, curiosity, desire for personal control and mastery, are therefore essential elements supporting the survival of genes, organisms and species alike (Buss, 1988).

Thus, I argue that humans, being organisms and as such partly genetically determined, are equipped with evolutionarily advantageous genetic risk taking programs.

Evolution, Cognitive Abilities, and Attitudes

Cooper (1987) even suggested that cognitive patterns and structures of thinking are evolutionary determined. Cooper (1987) argued that decision heuristics and subjective probability theories are the evolutionary product of the survival of those organisms which developed the necessary brain structure to apply exactly these heuristics. The mere survival of organisms using a certain type of cognitive structure made the organisms using this structure better equipped, and therefore increased the likelihood of offspring, possessing this special cognitive pattern, as opposed to the less successful ones.

Cooper (1987) postulated that most people follow the same basic decision strategies, as they all underwent similar selection procedures in their evolutionary history. According to this argument, even complex cognitive processes interact with genetic predispositions over a longer time span of development (Crawford & Anderson, 1989).

Loehlin, Willerman and Horn (1987; 1988), Plomin and Rende (1991) as well as Tesser (1993) gave extensive reviews on the genetic components, the developmental and the evolutionary aspects of attitude and behavior formation in humans. They argued that many, if not most of our attitudes, our creativity and many mental disorders as well as temperaments and behavioral patterns do have genetic contributions and thus underlie the same evolutionary pressure as physical features such as weight, height or blood group. Furthermore, concepts that have been typically ascribed to human beings only, such as self-awareness, self-conscience, language etc. are also found with increasing empirical support in our evolutionary relatives, namely chimpanzees, orang-utans and to a lesser degree in gorillas (Povinelli, 1993; see also Premack, 1976). In addition, Lykken, McGue, Tellegen and Bouchard (1992) and Lykken, Bouchard, McGue, and Tellegen (1993) found support in twin studies for genetic contributions to psychological features, such as aggressiveness, taste in food and clothing, choice of profession, hobbies, names and interests.

Genetic Personality Components Found in Twin Studies

Some further support for the genetic determination of personality types and risk taking tendencies comes from twin studies. Bouchard (1986) reviewed the role of twins throughout history and demonstrated their special roles on examples like Castor and Pollux, or Romulus and Remus. Furthermore, Bouchard (1986) gave an example of the Roman play "Menaechmi" by Plautus, describing the development of two identical twins, separated at age seven. Galton (1875, cited in Bouchard, 1984) also suggested to study twins to understand the difference between genetic and environmental influences. The reason for this interest lies in the identical genes of monozygotic twins. When comparing the differences and similarities in personality structures of monozygotic twins with those of dizygotic twins one can attribute the differences to genetic influences. Also, when identical twins are reared under very different environmental circumstances similar personality structures can be attributed to genetic disposition.

The "Texas Adoption Project (Horn, 1983) and the "Minnesota Twin Study of Twins Reared Apart" addressed this topic under controlled conditions, in contrast to some poorly controlled research, or the allegedly falsified research by Sir Cyril Burt (Bouchard, 1984; Hoffman, 1991). Bouchard (1984) described in great detail which psychological tests have been employed. Areas of mental abilities, information processing, and vocational inventories, as well as personality inventories, psychomotor tests and others more have been used to test the twins in their study. Bouchard (1984) confirmed findings reported by Nichols (1978) who summarized much of the literature on this topic.

Nichols (1978) reported considerable genetic influences on the personality traits Extraversion, Impulsiveness, Neuroticism, Socialization, Dominance, Masculinity, Conformity, and Flexibility. Most of these traits also play an important role in risk taking behavior (Zuckerman, 1979). Scarr, Webber, Weinberg and Wittig (1981) reported results pointing in the same direction in a study on adopted children, who showed more similarities to their natural parents than to their adoptive ones. Both studies reported only minimal correlations with respective environments (0.04 to 0.08). Similar findings have been reported by Lykken et al. (1992) who found far-reaching similarities in monozygotic twins separated early in their lives, but not in dyzigotic twins. Segal (1993) reported similar findings in relation to altruism, cooperation and competition.

Such results led Bouchard (1984) and Lykken et al. (1992) to argue against Mischel's (1981) statement that personality is ***largely*** determined by the immediate environment. On the other hand Hoffman (1991) argued that finding similarities in identical twins reared apart is no doubt a strong argument in favour of genetic contributions, while not finding similarities in siblings is no argument against the importance of environmental factors. Lykken and Tellegen's (1993) studies on mate selection differences in monozygotic twins also set clear limits to the influence of genetic patterns.

Hoffman (1991) argued, that the folowing factors determine and lead to marked differences in the development and expression of psychological characteristics:

- ordinal position
- age of the child when an event occurs
- gender
- physical appearance
- idiosyncratic experiences

Hoffman (1991) pointed out that the second child, for example, has already more experienced parents, receives less attention due to the presence of the older sibling and generally has not to *break the way* in fulfilling those wishes the older child had to argue and fight for. In relation to risk taking behavior and strategies these educational surroundings should lead to very different strategies. Gender differences and the consequences for social and physical risk taking have been discussed already.

Furthermore, as the mental capacities of a child slowly develop over time (Povinelli, 1993) it is obvious, that events, such as a divorce, change in homes, cities, etc. are perceived and attributed differently by children of age 3, 6 or 13. The influence of physical attractiveness in success in mate selection has been discussed. Physical risk taking, such as reaching goals through violent means, defending oneself or one's friends in fights, being quicker, stronger, smaller etc. does have significant consequences for the child's experiences, self-worth, and of course risk taking strategies and tendencies.

Finally, although siblings are raised in the same family, they do make individual and independent experiences, such as rewards, observations, illnesses, accidents. This argument holds true for monozygotic twins as well and is one of the strongest contributors to their individuality and the psychological and physical differences found between them (Hoffman, 1991).

A further approach to study genetic influences was proposed by Scarr and McCartney (1983) and Scarr et al. (1981). They stated that individuals who inherit traits, such as intelligence actually inherit a propensity to seek out intellectually stimulating environments. This contrasts with the assumption that the structure of the nervous system directly determines personality and abilities such as intelligence (McGue & Bouchard, 1988). Buss (1991) reviewed several recent studies showing that people actively select, manipulate, and restructure their environment according to their needs and their heritable strategies. They thus establish a link between their dispositions and circumstances most beneficial for these disposition.

Arvey, Bouchard, Segal and Abraham (1989) also suggested that intelligence may play a moderating role in the decision to choose jobs, and that the genetic disposition has an influence on job attitudes. Arvey et al. (1989) reported a heritability factor of approximately 30% for perceived job satisfaction, in a study on twins reared apart. This finding, if it can be replicated, would confirm to some extend Staw and Ross' (1985) assumption that biologically based traits predetermine social and work decisions.

Baker and Daniels (1990) reported that monozygotic siblings had a stronger link between experiences while growing up and adult personality patterns than dizygotic siblings. They suggested that genetic differences may alter the perception of events to some degree. These different perceptions then lead to different personalities. Tellegen, Lykken, Bouchard, Wilcox, Segal and Rich (1989) reported significantly larger inter-correlations for monozygotic, than for dizygotic twins on the *Aggression, Control, Harm Avoidance, Alienation, Stress Reaction*, and *Positive*, as well as *Negative Emotionality* sub-scales of the *Multidimensional Personality Questionnaire* (Tellegen et al., 1989). As variables, such as aggression, control, harm avoidance and stress reaction are obviously related to behavior in uncertain or risky situations, these reported findings further support the postulate that risk taking behavior is at least partly determined genetically.

Finally, Fulker, Eysenck and Zuckerman (1980) reported in a study of 422 twins, that sensation seeking, which is shown to be related to risk taking (Zuckerman, 1979) had a heritability coefficient of 58%, thereby providing strong support for a genetic influence on risk taking behavior. On the other hand, it is strongly suggested that educational and learning differences are major factors of influence in determining how one reacts in "risky" situations (Mischel, 1981; Scarr et al. 1981; Bouchard, 1984; Hoffman, 1991). If one was brought up in such a way that every source of uncertainty and thereby danger had to be avoided the person would developed a different cognitive set than the person that was brought up to enter the situation in order to master and ultimately control it. Research by Eron (1984), who conducted a 22 year-longitudinal study, and by Eron, Huesmann, Lefkowitz and Walder (1972), who studied the influence of television on aggressive behavior, showed evidence for the impact of early experiences on the development of aggressive behavior. Most studies attributed 50% - 70% of the variance to environmental influences.

Kagan and Snidman (1991) demonstrated that *inhibited* versus *uninhibited temperaments* in 4 months old babies were predictive of their anxiety and risk taking behavior in later years. They reviewed both animal and human research that hints at possible brain sites and physiological processes that may contribute to these relatively stable differences in temperament. They strongly pointed out that biological components to such differences in temperament do only change the probability of certain resulting behaviors to occur, but these dispositions should in no way be understood as inevitable or unchangeable, notions often wrongly associated with genetic patterns.

Buss (1984) argued for socio-biology to study the individual differences as well as stressing the human similarities. Buss (1984) pointed out that although evolutionary theory plays an important role in the development of human beings, changes in environment always occurred simultaneously, thereby making the distinction difficult. Furthermore, mass transportation, television, and mass media constantly increase the uniformity of populations (Buss, 1984), which can also confound heritability quota somewhat.

In addition, population changes, like assortive mating (Buss, 1983; 1984), may also account for finding certain personality traits more predominant in some groups. Buss (1983; 1984; 1984b) described the process of assortive mating as very important to the development of traits such as aggressiveness, venturesomeness, extraversion, and other intellectual and social factors. Mating with a partner of similar genetically controlled personality traits, will greatly increase the likelihood of these particular genes to survive and multiply. Buss (1984), however, also pointed out that environmental and situational factors can greatly influence the choice of one's mating partner, a notion strongly supported by Lykken and Tellegen (1993).

It seems to be important to combine evolutionary concepts with concepts of situational and individual differences to gain a full understanding of the relationship between risk taking and human personality (Turkheimer, 1991). Furthermore, focussing on only one particular statistical design of studies, such as correlational versus experimental design, will only widen the gap between what is studied and what should be known (Buss, 1984).

3.1.3 Risk Taking as Evolutionarily Advantageous Behavior

Such findings lead to the conclusion that during the course of evolution, attitudes, behavioral patterns and cognitive abilities have developed differently for different species, but that there are some underlying similar general patterns. These patterns seem to be rooted in the evolutionarily older areas of the brain and those closer linked to survival and to reproduction, as that are common components for all living organisms. Due to the development of societies and cultures, as well as large migratory processes, environmental changes and social developments, the base-levels for risk taking and risk-perception should vary widely within all present cultures and populations (Dawkins, 1976). Although our protective western cultures do not force people to take extensive risks to find food and shelter, the prototypes of a ***risk seeker*** (often closely related to stereotypes of a *"Hero"*) and a ***risk avoider*** (sometimes related to the stereotype of a *"Coward"*) are still very present, and because evolutionary changes take a long time, are not likely to disappear soon.

Furthermore, we protect risk avoiders and risk seekers alike through mandatory safety features. Societies also encourage risk taking, for example by giving "fame and fortune" to people risking their health and lives (e.g., race-car drivers, pilots, astronauts, boxers, explorers,), as well as through an economic system of free enterprise, which encourages financial risk taking, while social systems reduce some of the negative consequences resulting from errors in risk-assessment. Risk seeking tendencies are, therefore, not likely to become extinct (Crawford and Anderson, 1989). Within the frame of a protective society risk seekers are likely to find other means to satisfy their curiosity and to reach their optimal level of arousal, especially, as both processes continue to be intrinsically rewarded. It is therefore, likely that risk seekers will engage in activities allowing them to fulfill their desires, even within a protective society. Consequently, risk seekers should seek out professions that offer greater arousal and variability, seek out activities letting them experience a subjectively optimal level of risk and engage in behaviors allowing them to exercise "potential survival skills" in competitive play, and attempt to gain personal control over possible dangers by deliberate exposure to risks (Daly & Wilson, 1978; Wilson & Daly, 1985).

If risk-takers succeed in situations of danger, they are intrinsically rewarded for becoming optimally aroused, for reaching their goal and for gaining control and mastery. If at first they do not succeed in their task, however, their further action depends even more on the outcome of their behavior. If they are still able to correct the unsuccessful behavior they will most likely do so for two reasons. First, their action is partially reinforced by their very survival, and second, the risk of failing adds an additional arousal. It is, therefore, very likely, that people will continue to engage in subjectively dangerous activities.

It seems logical to assume, therefore, that risk taking is an essential behavior, which is probably partly programmed genetically into everyone. Thus, across the population as a whole, individual differences in risk taking tendencies will likely be distributed normally, namely in form of a bell-shaped curve. Such a distribution encompasses people who take risks always and everywhere, and those who never and nowhere take risks are very rare and at the extreme ends. The majority of people will be found around the center of the curve, as they constantly engage in a ***subjectively optimal*** degree of risk taking, since they are, presumably, intrinsically rewarded for exactly this behavior. The specific kind of risk taking, its characteristic features and qualities, however, may very well be distributed along different dimensions. Furthermore, individual evaluation, preferences, expectations and associations add to the variety of risk taking activities and experiences.

Zuckerman (1990) made a strong argument in favour of a normal distribution of most personality traits, including sensation seeking, criminality, psychopathy across the human races. According to Zuckerman there is no scientific basis to determine such racial differences as the methodology of twin studies, or adoption studies is not applicable. Thus, all allegedly found differences are not separable into their genetic, environmental and early developmental components. In addition, risk taking behavior has been found in all races and cultures with high degrees of variation both inter-individually as well as intraindividually across situations and time. Type and degree of subjective and objective risk taking, will therefore differ widely. Any behavioral error, resulting from a faulty risk assessment, or from a wrong action, can become very costly, physically, financially and socially. Following this line of argument, errors or accidents are not necessarily the result of a risk seeking desire, but more of inappropriate action, resulting in, or stemming from a faulty risk assessment.

Based on the postulate that risk taking is a genetically determined, an evolutionarily stable, a physiologically, emotionally and cognitively intrinsically supported and, furthermore, even extrinsically rewarded behavior, it is rather illogical to attempt to reduce errors and accidents by preventing people from such a highly rewarded activity as risk taking. The inevitable result will be that they search for other means of risk taking, replacing the one just prevented (Wilde, 1986).

In relation to risk taking, many attempts have been made to understand the underlying factors. Personality theories, which are largely based on biological trait models, have focussed on individual assessment, while decision making under uncertainty theories have attempted to clarify situational processes and moderating variables. Both approaches have, however, largely studied either cross-sectional samples in between-subject designs. Longitudinal, or within-subject studies are a scarcity.

Also, most researchers did not work interdisciplinarily, and disregarded emotional or other non-material factors and processes. Their results are, therefore, incomplete. However, as personality theories and situation-dependent theories are both valid and important sub-factors in the concept of risk, I shall discuss these approaches next.

I shall begin with biologically based arousal theories, as arousal can be considered an automatic by-product of experiencing risks, and appears to be closely related to evolution. Differences in arousability of the nervous system could be genetically controlled, and inherited, especially if assortive mating (Lesnik-Oberstein & Cohen, 1984) takes place. It may thus form the basis for different temperaments, personality traits, and behavior.

3.2 Risk Taking and Arousal

Based on the necessity for high degrees of physical and mental alertness in situations of risk or uncertainty these situations are likely to be accompanied by some kind of physiological and/or psychological arousal to ensure fight/flight preparation. Thus, to understand the processes connected to risk taking behavior it is important to incorporate the concept of arousal. The study of arousal, however, has evolved into such a large area that it would take a whole book to review it. I shall, therefore, discuss the research relevant to risk taking.

3.2.1 Historically Early Arousal Theories

The idea of an optimal level of arousal for maximum performance is very old. Hippocrates (400 B.C./1849, cited in Murray, 1988) discussed activation levels related to personality types, while the philosopher Bain (1875) described the relation of optimal level of arousal to emotional reactions. Furthermore, classic literature is full of examples of rages, power bursts and descriptions of extraordinary actions under conditions of threat (Illiad, Odyssey, La Morte d' Arthur) and of calm and quiet thinking as a means of solving cognitive problems (e.g., De Bello Gallico, Sherlock Holmes).

One of the first recorded scientific investigations of optimal level of arousal theories can be found in Wundt's (1874) experiments analyzing the relationship between stimulation of pressure, temperature, taste, and olfactory senses and feelings of pleasure and pain. Wundt (1874) expressed his findings in the form of an inverted-U curve, but also stressed that some senses, such as sweetness are always considered pleasant and that "higher" senses were not necessarily following the same pattern.

Freud and Breuer (1895/1955) assumed that organisms have a tendency to maintain a constant intracerebral excitement level, and that there is an optimum level of excitement, beyond which excitement becomes aversive. Too low an arousal level, such as in darkness or quietude, was also assumed to be agitating and unpleasant. (These assumptions were later supported by Zuckerman's (1979) studies on sensory deprivation.) In the same work, Freud and Breuer (1895/ 1955) suggested that profound differences of the nervous system are manifested in the energy set free by the functionally resting cerebral elements. This may lead to individual differences in the optimal level of arousal. Thus, early studies on arousal seemed to suggest that in order to face dangerous situations an increase in physiological arousal is desired.

3.2.2 Drive Reduction Theories

Later on, however, Freud changed his theory towards a drive-reduction approach, culminating in the formulation of the concept of a death instinct (drive): ***"The goal of all life is death"*** (Freud, 1920/1955).

The idea of drive-reduction or the desire to fulfill needs became very popular in the following decades. This assumption would mean that risk taking is a tolerated rather than desired or rewarded behavior. Drive reduction theories would argue that people only strive for a reduction in risk, and that they would not actively look for increases in risks.

McDougall (1923) followed Freud's (1915/1957) suggestion that innate instincts govern our behavior, while Woodworth (1918) used Freud's later concept of "Trieb", (= drive, often wrongly translated as instinct), to explain how activities, such as curiosity and sex, activate other mechanisms, such as habits (Zuckerman, 1979). Tolman (1926), Hull (1943), and Spence (1956) used drives and drive reductions as their key concepts, while Murray (1938) used the term "needs" to describe behaviors such as playing, exhibition, sex, sentience, change and many more.

Allport (1937) and later Maslow (1954) detached drives and needs from innate biological motives and postulated a developmental process involving several steps. The lower steps of physiological needs and safety are followed by needs for love, belongingness, esteem and finally self-actualization (Maslow, 1954). The last mentioned need cannot be fulfilled by simple biological instinctive processes, but depends to a large extent on experience, individuality and external situations. Self-actualizers seek out novel experience-producing feelings of aliveness and fascination (Maslow, 1968). This description is similar to Zuckerman's (1979) description of a sensation or risk seeker. These formulations, as well as findings by Harlow (1953), substituted a purely drive reductionist theory with other approaches, where the increase of arousal can also be experienced as pleasurable. Theories that assumed the experience of pleasure for both increasing and decreasing arousal were the optimal level of arousal theories.

The consequences of optimal level of arousal on risk taking are that the desire to take risks up to an optimal point as well as the reduction of risk beyond this optimal point would be accompanied by arousal and intrinsically rewarded. They can thus explain why people actively search out risks, a behavior drive-reductionists could not explain.

3.2.3 Inverted-U Theories of Optimal Arousal

Yerkes and Dodson (1908) discovered that performance in a complex task was related to arousal in a curvilinear fashion, namely in the shape of an inverted-U, while performance on simple tasks was linearly related to increasing arousal. The Yerkes-Dodson Law states that with increasing levels of arousal performance improves, until a peak performance is reached. Every additional stress or arousal will lower the level of performance in complex tasks, while performance in simple tasks will continue to increase for some more time before it also levels off and ultimately drops. That implies that too low arousal is just as bad for performance as too high arousal, because the performance detriments are equal on the left and right side of the curve.

Duffy (1932; 1957) applied the Yerkes-Dodson Law to her arousal theory splitting behavior into direction and energy mobilization, being roughly equal to Hull's concepts of habit and drive (Zuckerman, 1979). Duffy (1957) further suggested that individual differences are due to biological and environmental factors, which are reflected in the EEG. In the EEG, high arousal is represented by fast alpha rhythms, which are often found more pronounced in risk seeking individuals (Zuckerman, 1983).

Lindsley's (1957) theory of motivation assumed a thermostat function of the reticular formation, which regulates activation in a homeostatic fashion. Lindsley (1957) suggested that children's curiosity, playfulness, and hyperactivity may be a result of the gate function of the reticulo-cortical system, which gives preference to novel stimuli, blocks repetitive ones, and thus accounts for habituation (Zuckerman, 1979).

The early optimal level of stimulation theory by Hebb (1949) also assumed a drive reduction approach. Hebb (1949) stated that people seek out "pleasant fears" in certain activities, like riding roller coasters, and that fear becomes unpleasant at higher levels of arousal. According to Hebb (1949), seeking out fear was done to increase the arousal in order to experience the pleasurable feeling of arousal reduction.

Later on, however, Hebb (1955) changed his opinion due to physiological findings relating a necessary brain activation to thinking and conscious action, which implied that not only arousal reduction, but also arousal seeking was desired by the brain. In 1955, Hebb adopted the inverted-U shaped curve from the Yerkes-Dodson Law.

Berlyne (1960; 1971) developed the theory of "arousal jags" to account for topics, such as curiosity, exploration and aesthetic preferences. He combined experienced arousal with other affective and hedonistic experiences and incorporated findings by Olds and Olds (1965) regarding self-stimulation. Berlyne (1960) defined the specific determinants of arousal potential as: size, colour, intensity, sensory modality, affective connotations, novelty, complexity, degree of change, suddenness of change, surprisingness, incongruity, and uncertainty. If the arousal potential is unusually low, boredom occurs, if it is too high it is also experienced as unpleasant. Both states are accompanied by high levels of tonic arousal.

Similar to Hebb, Berlyne also shifted from drive reduction (Berlyne, 1960), to an optimal level of arousal approach that includes positive feelings resulting from increasing arousal (Berlyne, 1971). Based on findings by Olds and Olds (1965), that certain areas in the limbic system produce primary reward effects whereas other areas produce avoidance, Berlyne (1971) suggested different thresholds for the system. With the threshold of the avoidance system being higher than that of the primary reward system, Berlyne (1971) reproduced Wundt's (1874) inverted-U curve, as a function of the activity in the two limbic systems.

Another approach toward explaining arousal was employed by Thayer (1967; 1972; 1978; 1987). In his two-dimensional arousal model, Thayer (1987) differentiated between energetic and tense arousal. Energetic arousal is similar to Eysenck's (1967) cortical arousal in the reticular activation system. It is influenced by such factors as the sleep/wake cycle, including rest and sleep, and voluntary gross motor activity. Energetic arousal is assumed to play a role in information processing. Tense arousal, on the other hand, is primarily associated with dangers, real or imagined, and its components of emotional reactivity and anxiety correspond roughly to Eysenck's (1967) autonomous activation system (Thayer, 1988).

Thayer (1967; 1972) stressed the interaction between cognitive and physiological arousal and suggested to measure arousal through self-reports. He found stronger correlations between self-reported arousal and total physiological arousal, than between direct methods of measuring physiological arousal, such as GSR and heart rate variability.

The theories discussed so far provided support for the assumption that situations of danger and risk are accompanied by physiological arousal, and that people actively try to adjust the arousal level to an optimum. Cognitive and emotional aspects of risk taking behavior and arousal, however, have not been explored sufficiently.

3.2.4 Theories Relating Cognition, Arousal and Risk Taking

Fiske and Maddi (1961) used a broad concept of stimulation, including interoceptive stimulation and stimulation of cortical origin (thoughts or images). The stimuli varied in terms of intensity, meaningfulness, and variation of their influence on arousal and resulting behavior. Fiske and Maddi (1961) referred to activation as the excitation of the CNS, whereas arousal referred to peripheral effects, like muscular reactions. On detection of a deviation from optimal arousal, the organism compensates by actively seeking arousal increase through stimulation or reduction. The pleasantness of the experience depends on the degree of deviation from optimality, the complexity of the task and deviation of stimulation from the diurnal sleep-wake cycle. Individual differences and preferences were seen by Fiske and Maddi (1961) to stem from biological and early environmental differences, such as enriched or deprived sensory experiences. Fiske and Maddi (1961) strengthened the line of research on cognitive components of arousal, then further developed by Schachter's (1970; 1971) and Schachter and Singer's (1962) two-factor motivation theory. These authors claimed that unspecified arousal is first cognitively interpreted. Then an emotion is attributed to observation and experience.

The importance of cognitive processes and their relation to arousal was examined in studies of human error and accident causation due to mental overload (Hamilton, 1979). Easterbrook (1959) stated that emotional arousal progressively restricts the use of cues in the task environment. This leads to better performance in simple tasks, because distracting cues can be disregarded. It leads, however, to an impaired performance in complex tasks, since these tasks depend on a broader base of cues. With this explanation Easterbrook solved the circularity inherent in the Yerkes-Dodson Law, which measured arousal by its outcome, namely performance, even though performance was also the dependent variable.

Hockey (1979) assumed that there are several states of high activation. While the subject is expecting a stressor in a situation of risk or danger, all unnecessary bodily functions are reduced (e.g., digestive system) and the activation process remains in a static state. In the action or dynamic state, the body responses by mobilizing extra energies necessary to provide the resources for emergency action at high levels of effort. Hockey (1979) further supported Easterbrook's (1959) concept of focussing attention by reporting experiments where sleep loss (low arousal) decreased selectivity and noise (high arousal) increased it. In a study correlating EEG and drivers' behavior during prolonged driving under monotonous conditions subjects took greater risks when tired and focussed more on maintaining a constant high speed (Lemke, 1982).

These approaches can also explain the findings of Yerkes and Dodson (1908), namely that simple tasks improve and complex task suffer. Simple tasks need less attention or are automated, so they can be performed without much cognitive effort. If a stressor causes an alarm within a simple task situation, it is very likely, that the resulting focussing of attention will center even more on this task. Thus, performance will benefit from increased attention and physiological energy reserves.

If the task is complex, its interruption by an alarm will divert some needed mental capacity away from producing a solution. The automatic focussing process after arousal can compensate for this additional mental load, but only if the focussing centers on relevant cues for completing this task. A complex task, however, is likely to depend on broader perspectives and on several sub-tasks. A focussing on one of the minor components, or on a wrong solution, will therefore impair performance on the complex task.

The focussing of attention effect can also account for some contradicting findings in the literature. If a deterioration of performance did not occur, then the focussing may have been sufficient. If, however, performance on a simple task suffers, it may be due to either too much mental load by the stressor, or it may be due to focussing on minor or irrelevant aspects. It may also help to explain why in situations of uncertainty in simple tasks heightened arousal is desired (e.g., "psyching-up" for sports competitions) while it can be very detrimental in complex cognitive tasks ("blacking-out" in exams). The relationship between cognition, arousal and risk taking is thus more complex than assumed in earlier theories.

Empirical Support for Arousal Theories

Anderson (1990) summarized research in cognitive psychology that supports the inverted-U theories of arousal. Weltman et al. (1971) found support for Easterbrook's (1959) assumptions in that subjects paid less attention to peripheral stimuli, while perceiving themselves to be in a complex and dangerous situation, than when subjects did not perceive any risk. The performance of a simple task did not suffer.

Sperandio (1978) observed that air-controllers gradually shifted their attention to fewer and subjectively more important tasks and objects with increasing work-load. Under low stress, the controllers were able to detect many more errors in the peripheral field of their attention.

One well documented incidence of focussing on a minor aspect of a task while the performance of the task itself deteriorates is the "Air India" accident. An obviously defective control lamp that should have indicated whether the landing gear had been engaged, diverted the attention of all three crew-members who focussed all their attention on solving this minor problem. The primary task of keeping the plane in the air suffered in spite of warning signals. Although the flight indicator gave a clear auditory alarm of the plane losing height, no one of the crew noticed it until the plane crashed (*Why Planes Crash*, 1985).

The relation between performance and arousal holds just as well for ***perceived stressors***. For example, unreasonable panic is just as arousing as actual stressors (Baddley, 1972; Bainbridge, 1978). Baddley (1972) showed that SCUBA-divers under conditions of stress, namely suffering from nitrogen narcosis (depth sickness), were more impaired under water in a secondary task than in the pressure chamber. Baddley attributed this to the ***perceived*** higher danger under water.

Frankenhäuser (1980) stated that focussing on the danger to the exclusion of other present activities is not only a reminiscent of survival reactions. It also plays an important role in adjusting and responding to cognitive and emotional pressures of modern society. The power of subjective arousal becomes apparent in studies of intellectual efficiency, where it is well established that highly aroused people perform worse if presented with additional information, while persons with low arousal benefit from that even though the tasks are objectively judged as identical (Mandler, 1975). Furthermore, Holsti (1978) reported that highly aroused people found it hard to brainstorm, to give novel responses, to resist old categories, to perform complex response patterns, to delegate, and to resist information supporting their previous position (Weick, 1984).

Staw, Sandelands and Dutton (1981) reported similar findings in that highly aroused people who tried to cope with problems often reverted to more dominant, first learned actions, missed changes in stimuli that indicated new situations and forgot the latest learned behavior first. Weick (1984) concluded that heightened arousal can improve performance if it occurs after the focussing process has already occurred and only if the behavioral solution is well learned.

Fenz (1975) examined the arousal levels of novice, intermediate and expert parachuters. He found that the level of anxiety stayed the same, but that the time of arousal shifted to an earlier stage with increased experience. While the novices were most aroused when jumping out of the plane, intermediates showed the highest arousal when entering the plane and the experts were most aroused several hours before the jump or when packing the parachute. Thus, it was shown that expectations of danger can lead to as much arousal as the experience of the dangerous situation itself.

Similar arousal shifts were observed by O'Connor et al. (1985) who noticed that experienced soldiers had higher heart rates before, but lower ones during the jump. Ursin, Baade and Levine (1978) conducted an extensive study of young male military parachuters and examined the effects of biochemical, psychological and situational factors on jumping performance. They found clear relationships between activation, biochemical substrates and performance, as well as large individual differences. They strongly suggested to include all these factors into research on risk taking and arousal. McMillan and Rachman (1988) characterized parachuters as either courageous, over-confident or fearless, based upon a cluster analysis. All three personality-types differed in physiological responses to their first jumps, as well as in their observed behavior. McMillan and Rachman (1988) interpreted these individual differences as due to different coping mechanisms and arousability in the fight-flight reaction pattern.

Kaspar and Vesper (1976) reported that motorcyclists show significantly lower death anxiety than their peers. This was interpreted as the result of different coping or defense mechanisms.

An important feature of arousal, performance and risk taking that becomes explicit in these experiments is that arousal depends on the amount of skill required by the task. Novices show generally higher arousal or stress levels that are also often detrimental to their performance. Heightened arousal in relation with complex task can thus add to the increase in riskiness of a situation. The nervousness of young and inexperienced drivers can lead to more errors and a higher accident risk. Yet, high arousal due to the fear of being late (or in a car-race) can lead to increased performance on the driving task of experienced drivers, since the increased arousal helps to focus attention on the traffic. Expert drivers are not distracted by concentrating on how to handle the controls of the car.

3.2.5 Critical Points in Inverted-U Arousal Theories

Some critical points of the above named theories were summarized by Neiss (1988; 1990). Neiss (1988) reviewed the research dealing with psychologically induced arousal and its hypothesized inverted-U function relationship to performance. Neiss stated that the Yerkes-Dodson Law (1908) can only be applied in a trivial sense, as the concept is very broad and therefore not able to differentiate between optimal levels of arousal, and debilitating high or low levels. Furthermore, the concept cannot differentiate between specific types of arousals related to specific emotions, such as fear, sexuality, anger or joy.

Neiss (1988) further stated that the affective and cognitive components of arousal are highly interactive and can hardly be separated. Arousal, whether experimentally induced or occurring in real life, is always part of a larger psychological change. Neiss (1988) discussed the fear of an electric-shock. Arousal occurs here as a **consequence** of fear, and is therefore a possible contributor to performance decrements. However, because fear also causes other reactions and effects, (e.g., it diverts attention, Neiss 1988) the causal attribution of performance decrements to arousal is severely confounded. His review gives overwhelming evidence for the inconclusiveness of the effects of higher arousal, thereby indicating that other moderators may also, or instead influence behavior.

Further problems with the inverted-U theories included the circularity of arguments with arousal being effector and outcome simultaneously, and violations against basic requirements for theories, such as the impossibility to disprove it (Popper, 1974) are raised. No matter what the empirical findings, they can be accommodated within the theory, as every state of arousal could still be too far below the optimal level to increase performance (Duffy, 1962) or already beyond it, resulting in an identical performance. In addition, the task may be either too simple or too complex to be affected in the predicted way (Oxendine, 1970).

Furthermore, most studies seemed to show inverted-U results only in very threatening or other extremely high-arousal situations, whose inducements are under heavy ethical restraints. Neiss (1988; 1990) concluded, that all studies clearly show that arousal and performance are correlated, but nothing has been proven as to any causal relationship. Neiss (1988) suggested that we study instead the different physiological, emotional and cognitive states that lead to different behavior in different individuals. This analysis suggests that although arousal is clearly related to risk taking, and often in a curvilinear way, the exact relationship and the conditions that lead to it are unknown and possibly even experimentally impossible to prove. Theories of risk taking behavior and arousal must, therefore, include more complex measures than physiological levels of arousal alone. The influence of cognitive, emotional and task-dependent variables have to be considered.

3.2.6 Risk Taking, Arousal, and Physiological Measurements

A complete understanding of risk taking is only possible if the physiological processes underlying it are also considered. As arousal undoubtedly accompanies experiences of risk, and as the level of arousal may be important, efficient methods of assessing arousal have to be used. In the following, I shall discuss the physiological measures most often used in the study of risk taking and arousal.

Arousal can be split into psychological and physiological components (Singleton et al. 1971). Psychological arousal can be described as a feeling of excitement. As this feeling is often based on a physiological sensation of change within the central nervous system, psychological arousal is often dependent upon a conscious interpretation of a selected physiological arousal. The exact neural mechanisms involved in physiological arousal are described by Warburton (1979), and Sharit and Salvendy (1982). Because many measurements of arousal are physiological, I want to summarize the basic connections between neural mechanisms of arousal and behavior under high mental load or risk.

Arousal is controlled by cholinergic mechanisms via the ascending reticular activation system (ARAS). These pathways control the amount of cortical arousal at the sensory cortices. The interaction of the incoming evoked potentials with electric-cortical arousal reinforces the older components of the sensory evoked potential and constitutes stimulus selection. However, because not every random stimulus should be able to evoke electro-cortical arousal, a hippocampal feedback pathway to the tegmented area of the reticular formation inhibits too low-information stimuli in producing electro-cortical arousal. This process is important, because electro-cortical arousal is connected to the release of stress hormones. Corticosteroid release occurs in situations of high mental load, danger, or uncertainty. Adrenal catecholamines are also released in these situations. Because high-stress situations require a lot of energy, it is essential that these corticosteroids and catecholamines mobilize blood sugars and fats, which are the energy source for mental as well as physical action. According to Warburton (1979), this increased mobilization enables extra effort and high performance levels for a short time. In the long run, however, sustained release of sugars and fats leads to severe depletion and fatigue.

Anticipation of action can be as tiring as the action itself, since uncertainty releases stress hormones (Frankenhäuser & Johansson, 1986; Tomaka et al., 1993). Thus, being afraid of a future situation can lead to the same physiological experience of risk and arousal, as actually facing the situation does.

Nicholson, Hill, Borland and Krzanowski (1973), Nicholson, Hill, Borland and Ferres (1970) assessed the relation between cortical arousal, measured through Electro-Encephalogram (EEG), physiological arousal, measured through Electro-Cardiogram (ECG), and self-reported arousal during simulated flying. They found all three measures able to identify the degree of mental workload on the pilots, and that arousal was correlated with perceived dangerousness. Physiological measurements have been used successfully to assess the physiological correlates of risk taking, arousal, and mental load. For example, Wierwille (1979), Wierwille, Rahimi and Casali (1985), and Kramer, Sirevaag and Braune (1987) studied the physiological correlates of stress and risk in air-crews during simulated and real flights. Wickens (1979) examined the relation of physiological measures and secondary tasks and Wildervanck, Mulder and Michon (1978) studied the relation of physiological measures and mental load in driving. Other applications are listed in Wickens and Kramer (1985).

Heart Rate, and Heart Rate Variability (HRV)

One method often used to measure physiological arousal in risk taking lies in assessing activities of the cardiac system. Pribram and McGuinness (1975), Rabbitt (1979), and Van der Molen, Somsen, Jennings, Nieuwboer and Orlebeke (1987) successfully used heart rate as a correlate of arousal, or risk taking. Epstein and Roupenian (1970), as well as Thackrey and Pearson (1968) provided further support for heart rate as a relatively crude, but consistent and easily applied measure of arousal in relation to risk taking. Greater accuracy in detecting momentary changes in arousal, seems to lie in the measurement of heart rate variability (Mulder, 1979).

Aasman et al. (1987) reported the successful discrimination between input and mental effort by using heart rate variability. Takata (1982) performed a spectral-analysis of HRV, heart rate and respiration variability and found strong correlations with increased behavioral arousal and decreases in performance. Furedy (1987), as well as Jex and Clement (1979), however, reported shortcomings and possible confounds of the measure. Vincente et al. (1987) suggested to use the T - Wave amplitude change instead, but have not yet provided sufficient empirical evidence for the superiority of this method. Miller and Ditto (1989) showed that subjects' heart reactivity was inversely related to a video game penalized with electro-shock and that there was a significant influence of individual differences in relation to heart reactivity.

Galvanic Skin Response (GSR)

Studies related to risk taking and driving have often used the measurement of the Galvanic Skin Response (GSR) (Epstein & Roupenian, 1970). The method measures the skin's resistance to electrical currents between two electrodes. As the sympathetic innervation system becomes aroused or stressed, the subject transpires and this increased humidity changes the conductance of the skin. Through electrodes on the skin the specific pattern of these changes in resistance can be recorded and then correlated within time or space (Taylor, 1964) to changes in risk taking, or arousal.

Hulbert (1957), Michaels (1969), and Taylor (1964), found significant correlations of GSRs and the location of traffic accidents having happened there before. This finding may be partly due to prior knowledge, danger-indicating clues, or increased traffic at the spots.

Brown and Huffman (1972) distinguished drivers with good records from drivers with bad driving records via GSR, while Grant (1985) reported significant correlations between subjective risk-ratings and GSRs while driving an instrumented car. Näätänen and Summala (1976), and Boucsien and Hoffmann (1979) reviewed the literature and concluded that GSR can generally serve as a good instrument to assess subjective risk. In contrast to these findings Heino, Van der Molen and Wilde (1990; 1992) could only partly support this notion. Furthermore, individual differences in sensation seeking produced significantly different physiological responses. The use of

GSR and HRV as general and reliable methods of measurement for perceived risks during driving has to be considered with some caution.

Biochemical Analyses

Every electric activity of the brain is accompanied and performed by chemical substances that are either transported through the blood or connect synapses at nerve-links. As arousal is closely related to the adrenerg - sympathetic system, it is assumed that in states of higher mental stress more of the related substances will be used and therefore more of them should appear in the blood and urine of the subject (see also Zuckerman, 1979; 1983; 1985; Ursin et al., 1978; Stein, 1983). Warburton (1979) cited further studies showing that an increase of arousal was accompanied by increased secretion of corticosteroids (ACTH), as well as increased release of adrenal medullary steroids (catecholamines). Frankenhäuser and Johansson (1986) could show that increased adrenalin release correlated very highly with excessive overtime of assembly line workers.

3.2.7 Problems with Physiological Measurements

All cardiac measurements suffer from the shortcoming that an increase of activity can not only be attributed to an increased arousal, but also to mental load, or it may be an artefact of physical movements (Strasser, 1979; Heino et al., 1990). Biochemical measurements are prone to be constituted, and confounded, by physical and mental load, as well as by normal bodily reactions. Unless more specific substances can be linked to risk taking, biochemical measurements can only roughly differentiate between generally higher and lower levels of arousal.

Wierwille et al. (1985) assessed 16 measures of the mental load to pilots during flight and found only the eyeblinks and fixation fractions to be related to stress. It has been well recorded (Strasser, 1979), that physiological measurements in general suffer from great individual differences and different people may use different pathways in the CNS (Vanderwolf & Robinson, 1981).

Strasser (1979) listed studies in which one type of physiological measurement worked for one subject, but not for the next one. Before employing a physiological measurement, it is therefore extremely important to determine individual baselines and sensitivities to the various measures. Even if such precautions are taken, however, Blascovitch et al. (1992) showed that the relationship between cardiac arousal, affect intensity and self assessed intensity are either unrelated or inconsistently related to each other.

3.2.8 Psycho-Physiological Coping Mechanisms

Organisms can not maintain a level of extreme alert for extended times without severe energy loss. Being mentally alert, having all muscles ready for immediate and extremely quick actions and responding quickly to any possible danger do cost a lot of energy. Baum and Fleming (1993) showed that especially stress related to accidents and toxic hazards may lead to long-term chronic illness and mental problems. It is, therefore, most likely that mechanisms exist which neutralize the stress or danger reaction. These mechanisms should be automatically setting in when dangerous physiological depletion or saturation values are reached. Psychologically, the coping mechanisms should make the formation and utilization of stress and arousal controlling thoughts, emotions and behavior easier.

Physiological Mechanisms Changing Levels of Arousal

Solomon (1980) assumed a physiological process that takes effect after the stressor has raised the arousal level. Psychological mechanisms set in which reduce the negative effects of stressors and unpleasantly high arousal. Coping strategies, such as interpreting stressors as a challenge, rather than as a threat are discussed, for example by Dienstbier (1989) or Tomaka, Blascovitch, Kelsey, and Leitten (1993).

Opponent process theory, as put forward by Solomon (1980) builds upon work by Beebe-Center (1932), and Solomon and Corbit (1974). These approaches postulated the existence of two mutually counteracting processes that interact to produce balanced emotional responses (Mauro, 1988). According to opponent process theory, mammal brains are structured in such a fashion as to oppose any large emotional divergence from neutrality, whether pleasurable or not (Solomon, 1980).

Using data from Epstein (1967) on military parachutists Solomon (1980) gave an example of how the opponent process may work in a real-life situation: During the first free-fall, before the parachute opens, even the bravest person shows a fearful reaction. Their eyes may bulge, their lips may retract, they may yell, curl forward and their autonomous nervous system is in a highly excited state. After they have landed they go through a short period of looking stunned, not talking much and being stone-faced. Then, after a few minutes they begin to engage in lively interaction with their comrades. The after-reactions stay for at least ten minutes. After many parachute jumps the aftereffect is clearly visible, while the fearful reaction remains largely undetectable. Instead, the parachutist looks tense, eager, or excited, and during the free-fall reports a "thrill". After a safe landing there is evidence of a withdrawal syndrome. The activity level is very high, with leaping, shouting, and general euphoria. This period, often described as exhilaration, decreases slowly in time, but often lasts for two-three hours. A new, positive source of reinforcement is now available, but one that could never have occurred without repeated self-exposures to initially frightening situations the subject then becomes accustomed with.

To follow up on the example of the parachutist, Solomon (1980) stated that through the process of habituation to the initial fear stimulus the emotion of fear will be reduced over time, while the opponent process of joy will be strengthened by forming a new, pleasurable motive, reinforced by the pleasurable withdrawal syndromes (Solomon, 1980). Sports that include physical injuries, pain and possible death, would therefore not only be performed for the possible economic or social side-benefits, but also for the sake of experiencing this opponent process of a positive emotion.

Therefore, Solomon (1980) postulated that an innate opponent process maintains a stable emotional state or compensates for excessive emotions through opposite emotional responses. The assumption was based on research on the imprinting process (Lorenz, 1935) in ducklings (Hoffman & Solomon, 1974), on shock avoidance in rats (Maier, Rappaport & Wheatley, 1976), and aversive conditioning in dogs (Katcher et al. 1969).

Opponent process theory has been used to explain a variety of human behavior, such as opiate (Solomon, 1980) and cigarette (Pomerleau, 1979) addictions, students' responses to test taking (Craig & Siegel, 1979) and mothers' responses to breast-feeding (Meyers & Siegel, 1985). According to Mauro (1988), these studies suggest that fear, joy and distress may elicit affective after-reactions, but sadness does not.

Studies by Piliavin, Callero and Evans (1982) and Ranieri and Zeiss (1984), however, could not show an opponent emotional reaction to appear. Most of the studies in favour of opponent process theory, however, did not rule out other possible explanations for the occurrence of opposite affective reactions (Mauro, 1988).

Opponent process theory can be criticized on a number of different points. The hypothesis of delayed activation of the arousal reduction process by an arousal processes seems problematic when one considers direct activation of the arousal reduction-process and its assumed automatic effect on an arousal process, which has not even taken place. Which emotions result?

Furthermore, the model does not explain why and how the maximum size of arousal-process sometimes, but not always, exceeds the size of the arousal reduction-process and how the timing is processed in just the exact time for an under/overshoot reaction, that then levels out. It is not a dynamic process model, but was rather developed to fit existing data (Grossberg & Gutowski, 1987).

The concept of personality differences in responsivity, interpretation and activity is not clearly discussed in Solomon's (1980) paper. But one can assume, that different learning, reinforcement, imprinting or situational changes would lead to the formation of different stimulus-response patterns, that then generalize to new situations. Take for example the parachutist who breaks a leg during his first jump. His opponent process of joy after fear will most likely trigger even more fear before the next jump and may very well end the parachuting career of that person. Other experiences and reinforced opponent processes to learned stimulus-response patterns may also prevent a person from ever trying any voluntary risk taking activity.

The findings on opponent process theory are not always conclusive, sometimes contradictory, and many of the involved factors have not yet been studied (Solomon, 1980; Murgo, 1988). Further examinations of different stimuli, different emotions and the role of individual reactions are necessary to clarify the theory, which so far has largely dealt with animal behavior. Nevertheless, Solomon's theory may be an important aspect of risk taking behavior since it offers an explanation for the finding that we do not stay at a high arousal level for an unlimited amount of time, but that automatic physiological processes reestablish a desired balance.

Psychological Mechanisms to Change Undesired Arousal

Physiological reductions of undesired arousal can be accompanied by psychological arousal-reduction processes. If the perceived risk-related arousal is too high, coping mechanisms to reduce either exposure to the stressor, or to allow a cognitive re-interpretation of the stressor take action (Dienstbier, 1989). Folkman and Lazarus (1988) pointed out that emotions can work as a mediator of coping strategies in response to negatively experienced stress, but that coping can also be effective as a mediator for the emotional experience of stress. If subjects had used an effective coping strategy for dealing with the stressor, they reported significantly less negative emotions than if they did not have a coping strategy.

McGrath (1976) and Kobasa (1979) also found higher levels of reported stress with higher levels of uncertainty. Kobasa (1979; 1982), and Kobasa, Maddi and Kahn (1982) interpreted coping processes in the light of personal hardiness, while Weick (1984) suggested that people compensate for detrimental effects of high arousal by applying a strategy of "small wins". The strategy suggests to split a complex problem into simple ones, which are then addressed individually. The approach reduces risks, arousal and makes it easier to utilize existing skills (Weick, 1984).

Dienstbier (1989) went one step further and described conditions when subjects experience stress and risks as positive, and even engage in an active search for arousal producing stressors. Dienstbier referred to studies on animals, that showed that those animals accustomed to stress, such as touching and electroshock, subsequently show less arousal in new experimental conditions imposing threats on them.

Blascovitch et al. (1993) supported Dienstbier's assumptions in a study examining the effects of induced active (solving mathematical problems) or passive stress (viewing a film) on physiological and self reported stress. They found that stressors perceived as challenge heightened cardiac reactivity and decreased vascular restriction. They interpreted this finding as indicative for a process of "energy mobilization" (see also Brener, 1987) for upcoming uncertainty. Subjects perceiving the situation as threatening, however, showed the opposite response patterns and reported higher levels of stress and lower levels of perceived control. These people are, according to Dienstbier's concept of "toughness", slower in physiological responses and less able to cope with novel or uncertain situations and demands.

Seligman and Weiss (1980) and Weiss et al. (1975) found that people interpreting stressors as a challenge, showed their "toughness" to an increase in the capacity to produce and store catecholamines, which are assumed to be influential in the reaction to threats, arousal and dangers. Similar relations between catecholamines, Monoaminooxidase (MAO) and risk seeking in humans were discussed by Zuckerman (1979; 1983; 1984).

Further support for a "toughening" ability of catecholamines was reported by Johansson and Frankenhäuser (1973) for increased resistance to errors in tests, and Frankenhäuser, Lundberg and Forsman (1980) in relation to achievement oriented situations for Type A and Type B individuals. Ursin et al. (1978) reported catecholamine increases after parachute jumps, and correlation of those catecholamines with performance measures.

Dienstbier (1989) suggested that the immune system of an organism can be improved by engaging in active "toughening" against catecholamine depletion through aerobic exercise and controlled exposure to threatening stimuli, or through passive "toughening" by exposure to coldness.

Allred and Smith (1989) found that hardy persons interpreted cognitively threatening situations as challenging, reported more positive self-statements and showed lower physiological arousal. Dienstbier (1989), Frankenhäuser and Johansson (1986) concluded that individuals who experience stress as a challenge respond with different hormonal reactions and experience positive feelings, and establish a resistance or toughness against future threats.

Thus, such individuals are more likely to seek out situations leading to positive risk-experiences and to further toughening. Individuals who experience stressors as a threat have different hormonal, emotional and behavioral responses, as they avoid situations that may expose them to threat, thereby preventing themselves from establishing resistance against future stressors.

Therefore, if risks and the resulting arousal are interpreted as a mere challenge, or accompanied by positive feelings, people actively seek out risky situations. If the risk they perceive is undesired, cognitive reinterpretation can reduce the level of arousal. This psychological process together with an automatic physiological arousal-reduction process could be effective in preventing the organism from being damaged by constantly too high arousal levels. Individual differences in the physiological and psychological propensities may also influence the perception and handling of risk taking related arousal and may determine what levels of risk are perceived as undesirably high.

3.3 Individual (Personality) Differences and Risk Taking

Individual differences in both the nervous system as well as in the personality pattern may determine whether a given risk is interpreted as a threat, as a challenge, or not at all. It may further determine whether the risk, once experienced, is undesirably high or low. One possible cause for such individual differences lies in a different structure of the nervous system.

3.3.1 Strength of the Nervous System Theories

Pavlov's Approach

One of the basic assumptions in the strength of the nervous system theories is, that there are biologically based personality types, based upon the concept of: sanguine, choleric, phlegmatic and melancholic personalities. The first of these "modern" personality theories was developed by Pavlov (1927/1960) and later refined by Nebylitsyn (1972) and Strelau (1974; 1982). The four personality types are based on doctrines of Hippocrates (about 400 B.C., 1849; cited from Murray, 1988), Galen (about 160 A.D.; *On the Passions and Errors of the Soul*, 1963), which were rediscovered and applied to modern psychology by Wundt (1874, *Grundzüge der Physiologischen Psychologie*). Pavlov (1927; 1955) described the individual differences of the dogs he used in his experiments on the basis of these four personality types. Each type referred to different behaviors of the dogs, like being active or passive, curious or indifferent. Pavlov (1927/1960, p.286) stated about "sanguine" dogs: "In dogs of this type many conditioned reflexes must be developed concurrently and with great variety of stimuli; no stimulus must be repeated more than once in a single experiment and long pauses between applications of the various stimuli must be avoided". Zuckerman (1979) called this type of dog, or person, "sensation seeker", who is looking for the optimal level of arousal.

Pavlov (1955) and his student Teplov called it the ***strength of the nervous system theory***, as they assumed that the "strong" and "weak" excitatory and inhibitory neurons determined the character of dogs and people alike (Zuckerman, 1979). Later developments by Hull (1943), Eysenck (1947), and Gray (1964) opened the field for the modern strength of the nervous system theories by Strelau (1973) and Gray et al. (1983).

Strelau's (1974; 1982) concept of different temperaments and "reactivity" of individuals, as well as Gray's (1964; 1972; 1983) biological two-process theory are firmly routed in Pavlov's (1927) approach. Pavlov's (1955) concept of individual differences was solely based upon CNS arousability and did not account for any psychological influences. It was supplemented by the influence of expectations subjects had about events (Gray, 1972).

Thus, an individual reaction to a dangerous situation is determined by the subjectively experienced sensation intensity. The subsequent response intensity is determined by factors, such as the stimulus capacity, the actual level of arousal and other individual differences (Strelau, 1982). The response intensity differences depend on the individual's arousability (Gray, 1964), the strength of the excitation process (Nebylitsyn, 1972), the sensitivity of receptors (Fiske & Maddi, 1961) and the sensitivity of the effectors (Strelau, 1974).

Intra-individual differences, for example, as a consequence of different optimal levels of arousal, are also common (Strelau, 1982). As causal agents several systems are discussed by Kretschmer (1944), who assumed the endocrine system to be responsible for sensitivity, energy, and for the speed of mental activities. Pavlov and his students argued that the cortex is solely responsible for differences in strength of the nervous system, whereas Eysenck (1967) assumed that the reticular formation cortex-loop is responsible for the occurrence of the Extraversion-Introversion personalities.

Gray (1972) demonstrated the significance of the limbic system for Extraversion, Introversion and Neuroticism. Zuckerman (1979) focussed on the limbic system, and the neuro-hormones operating in it, such as: monoaminooxidase (MAO), catecholamines, norepinephrines, and dopamins. Zuckerman (1983; 1985) reviewed the neurological processes involved in Sensation Seeking and risk taking.

Strelau's and Gray's Physiological Personality Models

Strelau (1974, 1982) attempted to combine endocrine, automatic and cortical activities, and named the biologically based individual differences ***temperaments***. Such temperaments interact with physical and social environment factors to establish the personality of a person (Strelau, 1982). Strelau (1982) stressed that biological factors underlie the evolutionary process and are more important in determining the personality type, than either the social or physical environment are. Klonowicz, Zawadzka and Zawadzki (1987) supported Strelau's (1982) suggestions by showing that highly reactive temperament types considered stressful tasks with more curiosity and positive affects, both in anticipating outcomes and in retrospectively evaluating them, than subjects low in reactive temperament.

Gray (1964; 1972; Gray et al., 1983) combined the Pavlovian line of biological data with psychological theories, like that of Eysenck (1947) (Zuckerman, 1979). Gray (1972) offered neuro-physiological support for three real nervous system structures as a basis for individual differences. Two of these personality dimensions, extraversion and introversion are orthogonal to each other. In this system, the stable extravert-neurotic introvert diagonal is called "anxiety" and depends upon activity in the septo-hippocampal stop system. The other diagonal, stable introvert-neurotic extravert is called "impulsivity" and depends upon activity in the septal-lateral hypothalamus approach system. Psychopaths, and criminals in general, would be found at the extremes, while other people are found somewhere along these dimensions (Zuckerman, 1979).

The third personality trait, the amygdaloid fight-flight system, is closely related to Eysenck's (1967) concept of psychoticism. The amygdala is also known to play a major role in the execution of defense-attack and escape behavior. Removal of the amygdala results in the "Kluver-Bucy Syndrome", which produces a confusion concerning the reinforcing quality of stimuli in animals, and the loss of fear of other animals and dangerous things (Zuckerman, 1979).

In spite of some support for Strelau's or Gray's theory (discussed in Zuckerman, 1979; 1983) there are, however, points of critique to be raised. The approach towards novel stimuli, the orienting reflex and exploration behavior are all determined by the same structures, which also mediate punishment and reward. This, however, is not necessarily true, as Stein (1983) pointed out. Furthermore, fight-flight behavior is also determined by one structure only, which indicates, that personalities strong in this structure should fight and flee at the same time. Yet, aggressiveness and cowardice are normally not linked (Zuckerman, 1979).

In Gray's (1972) theory, arousal was a general construct, with no hedonistic or reward-punishment dimension. This shortcoming has been argued against by Roviaro and Holmes (1980) and McClelland (1985). Finally, most of these biologically based personality theories did not measure "reactivity", "nervous system strength", or "temperaments" directly, but they only measured correlates that may have been influenced by a large number of agents other than the stimuli given. Cognitive interpretations, emotions, memories or wrong perceptions may have further altered the measurements.

A strength of the nervous system approach suggests, that high risk-takers have a strong nervous system, need high levels of arousal and are low in reactivity to threatening stimuli. Whether the high propensity to take risks is based upon differences in the nervous systems, or whether the nervous system adapts to risky activities of the individual has not been determined yet.

3.3.2 Eysenck's Personality Theory

Building upon work by Pavlov and Hull, Eysenck (1947) increased the influence of psychological factors. He outlined a hierarchical theory of traits, starting with habits at the basic behavioral level. Habits are grouped into hypothetical traits and traits are organized into types, which are classified into three underlying personality dimensions. Eysenck (1947) called these dimensions: Introversion-extraversion, neuroticism (emotional stability-instability) and psychoticism (tough-minded asocial tendencies-socialized humaneness) (Zuckerman, 1979). The basic factorial structure has been replicated in a confirmatory factor analysis by Zuckerman, Kuhlman and Camac (1988), in a study on 3810 pairs of Australian twins (Heath et al., 1989). Risk-takers would score high on the extraversion-scale since they are interested in new and novel experiences and would also score higher on the psychoticism scale since anti-social behavior is considered risk taking (Zuckerman, 1979).

Based on twin studies Eysenck (1952) argued for a strong genetic influence on personality types. Similar to Pavlov's "sanguine" dogs, Eysenck's (1957) extravert personalities were harder to condition under partial reinforcement, weak unconditioned stimuli and long conditioned-unconditioned stimulus time lags (Eysenck & Levey, 1972). They were always seeking new experiences, were active, and friendly. This behavior was attributed to an excess of CNS inhibition, rather than to high excitation. Bristow (1981) found significant correlations of the Psychoticism scale and extrapunitiveness, supporting the assumption that people scoring high on this scale are hard to condition and respond poorly to punishment. In 1967, Eysenck applied the Yerkes-Dodson (1908) law and proposed that extroverts have a higher optimal level of stimulation than introverts. This difference was attributed to differential thresholds in the various parts of the ascending reticular activation system (Eysenck, 1967). The extravert personality corresponded with Pavlov's "strong" nervous system type, and the introvert with the "weak" type. The nervous system of the introvert is much more sensitive to stimulation, so that higher levels of arousal are perceived as unpleasant. Extroverts, on the other hand, need higher arousal levels to function optimally, due to their robustness, or relative insensitivity towards stimulation (Eysenck, 1967).

Gilliland (1980) was able to support this assumption by finding an interactive effect of caffeine as a stimulant, arousal and verbal performance. Extroverts continued to improve their performance under high caffeine induced arousal, while the performance of introverts suffered. Eysenck (1967) attributed this sensitivity, and the resulting behavioral expressions of extra-introversion, to genetic factors, which are therefore considered to be hereditary (Eysenck, 1967).

Eysenck (1967) extended his assumption of hereditary personality types towards neuroticism, psychoticism, and to components of these higher order dimensions, namely, impulsivity, sensation seeking, and venturesomeness. Eysenck, Eysenck and Barrett (1985) developed an independent scale including impulsivity, after this concept had gained more importance in recent studies (Humphreys & Revelle, 1984). Similarities across age, countries and cultures were found by Eysenck and Long (1986). Although Eysenck and Eysenck (1976) claimed a genetic determination of psychoses, psychopathy, criminality, low empathy, non-conformity and low socialization, Eysenck (1967) originally stressed that extreme environmental influences can outweigh a genetic disposition. This allowed for learning processes and situational influences in the personality model.

Humphreys and Revelle (1984) reviewed studies on the introversion-extraversion concept and stated that it has been a very robust concept, stable across conditions, investigators and across types of measurement. They also suggested that the impulsivity component of Eysenck's concept is closer related to arousability than the other dimensions. Campbell (1983), Amelang and Breit (1983), and Gray (1972) attributed many of the effects of psychoticism and neuroticism to the influence of impulsivity.

Rawlings (1984) reported correlations of Eysenck and Eysenck's (1975) psychoticism scale with measures of impulsivity and a choice reaction time test, as well as a creativity test, indicating a connection between arousal seeking and creativity. Meyer and Shack (1989) connected the concept of extraversion with positive mood and the concept of neuroticism with negative mood. They assumed these converged concepts to be the two major dimensions of personality, with sub-dimensions in impulsivity and risk taking. These assumptions were empirically supported by Larsen and Ketelaar (1991).

Some further support for a relationship between Eysenck's assumptions about extroverts and anti-social risk taking was given by Vestewig (1977), who showed that extroverts preferred more risky options in a study with examples based upon "Portfolio Theory" (Coombs, 1972; 1975). Furthermore, a relationship between higher extraversion and more risk-inclined attitudes to money, saving, social security, taxation and gambling was found by Furnham (1987). Kozlowski (1977) reported that subjects characterized as "stimulation demanding" by Eysenck's and Strelau's personality questionnaires chose significantly higher risks in gambling.

Bullock and Gilliland (1993) examined whether Eysenck's biological assumptions about differences in reaction times, arousability, electroencephalic responses and individual responsivity and performance after caffeine intake between introverts and extraverts occurred under controlled experimental conditions. They found no significant differences in reaction time or self reported arousal. The EEG recording of the brainstem auditory evoked response (BAER) showed shorter latencies of Wave V and faster conduction times between Waves I-V in introverts. They argued that this offers some support for Eysenck's postulate that introverts are faster transmitters and are easier arousable in the CNS, although the authors also offer another explanation based upon the olivochochlear bundle. They concluded, however, that introverts may respond with higher levels of arousal in identical situations than extroverts. Their arousal was thus beyond the optimal level of arousal according to the Yerkes-Dodson (1908) arousal-performance inverted-U function.

Stelmack, Houlihan and McGarry-Roberts (1993) also found support for the hypothesis that extraverts have shorter movement times and p300 components of the brainwaves, indicative of quicker information processing, but also potentially influenced by different arousal levels. The relationship between sub-optimal, optimal, and beyond-optimal arousal levels and risk taking, or sensation seeking was extensively researched by Zuckerman (1979), whose approach will be presented later. Eysenck's view on psychoticism has had some impact on the relationship between personality patterns and criminal risk taking behavior, such as violence, aggressiveness, anti-social behavior etc., which will be discussed next.

3.3.3 Personality Theories and Criminal Risk Taking

Thornton (1985) defined criminal behavior as a form of risk taking. High-risk offenders were shown to have both high risk-preferences and a low tendency to evaluate risks before acting. Howard et al. (1984), and Widom (1977) reported psychopaths to score higher on the extraversion-neuroticism-psychoticism scale, thus partly supporting Eysenck's (1967) assumption of criminality having some personality trait basis. Widom (1989; 1991) however, largely attributed her findings to environmental influences, such as parental modelling. Also, in contrast to Dilalla and Gottesman (1991), Hare (1982) reported no relationship between psychopathic prisoners and the extraversion, or neuroticism dimension, but a significant correlation with the psychoticism scale. He found no relationship between psychoticism and psychopathy. The differences in results may be due to different methods of assessment (Hare, 1982) or differently sampled population subgroups. This problem is common to studies on criminal risk taking and genetics even in well-documented adoption research, such as that of Mednick et al. (1987), or Oliver (1985; 1988) (Widom, 1991).

Hare, (1970; 1980; 1982), Farley (1986), Lykken (1982), and Schalling (1978) compared the typical behavior of psychopaths with that of highly impulsive, or neurotically extraverted, or "low arousal types". Hare (1970; 1982) also reported that psychopaths responded slower or weaker to stimuli threatening pain or punishment. Mednick and Hutchins (1978) stated that psychopaths returned slower to base-rate, after having been aroused. They suggested a strong link between genetics, biology and criminal behavior. Eysenck (1967) also postulated a relation of genetic predisposition of arousability, which then influences the likelihood of engaging in psychopathic and criminal actions. Farley and Farley (1972) pointed out that sub-optimal arousal types engage more likely in heavy drug and alcohol consumption.

Dilalla and Gottesman (1991) summarized some of the literature on genetic influences in antisocial, violent, aggressive, and general criminal behavior and their physiological measurements (e.g. levels of testosteron, insulin, 5HIAA). They concluded that there is considerable support for the assumption that these behaviors do possess a genetic contribution. The following behavior patterns have been linked to criminal behavior (see Mednick et al., 1987; Fishbein, 1990): resistance to punishment, poor academic performance, impulsiveness, childhood hyperactivity, recreational drug use, preference for active social interactions, preference for broad ranging sexual experiences, risk taking and sensation seeking. The conclusion, however, that this relationship is due to genetic differences has found strong criticism.

Criticism of the Personality-Criminality Link

Blackburn (1975), for example, argued that criminal people are just more unpredictable in their arousal patterns, while Matthews (1985) demanded a more detailed analysis of the arousal-personality link as a whole. Blackburn (1979) reported that primary and secondary psychopaths responded exactly the opposite way than predicted by Eysenck's (1970), or Hare's (1970) models. Blackburn (1975) attributed many of the observed findings to labelling, rather than genetic factors. When people are identified as psychopaths, or criminals, or sociopaths, they identify with this label and show the typical behavior patterns. Rather than a genetic predisposition, a societal stigma is responsible for individuals to fall into these categories.

Matthews (1985; 1986; 1987) reported that the correlations between arousal, extraversion, noise, intelligence and time of day can only partly be interpreted with the inverted-U curve theory. The interaction between arousal and performance followed predictions of this law while the interaction between impulsivity and extraversion did not. The triple interaction produced results that could not be explained by any of the tested theories.

Poulton's (1970) statement that extroverts have a lower arousal level and therefore seek arousal could be confirmed in Matthews (1985), but not in

Matthews (1986), thereby further indicating the instability of the arousal-psychopathy relation.

Furthermore, Zuckerman et al. (1988; 1991) reported several studies factor-analyzing data collected with Eysenck and Eysenck's (1975) personality questionnaire, most of which came up with different factor structures, ranging from 3 to 6 factors. Since all structures have some degree of reliability, the question of validity of the concept as a whole arises.

Bishop (1977) and Block (1977) challenged Eysenck's whole concept of psychoticism. Heath and Martin (1990) found that the psychoticism factor consists of a genetic and an environmental component which show opposite correlations with tough-mindedness and hostility to others. These findings are contrary to predictions made by Eysenck and Eysenck (1976).

Smith, Wilson and Jones (1983) suggested that impulsivity, and not extra-version influences physiological measure, while Thayer (1967, 1972; 1978) reported large individual differences in physiological measures of arousal in reaction to heightened stimulation. Thayer (1967; 1972; 1978; 1987) suggested using verbal self-reports of arousal instead, as they showed higher overall correlations to arousal than any individual physiological measure.

In an intensive study to test predictions from Eysenck et al's (1985) arousability concept, Stemmler and Meinhardt (1990) found that 48 experimental conditions and 34 physiological measures tested showed results contradicting assumptions made by Eysenck et al. (1985). Zinbarg and Revelle (1989) found that high anxiety hindered learning and conditioning more in "high impulsive" subjects than in "low impulsive" ones. This result is inconsistent with both Eysenck and Eysenck's (1976) as well as with Gray's (1972) predictions.

Widom (1989; 1991) showed that most empirical data on that topic, especially those on child abuse can be readily explained without using genetic concepts, that the methods, samples and conclusions are either flawed or too strong in conclusions for the data gathered. Overall, however, Widom (1991) agreed, that there is some support for a biological basis for the traits of extraversion and introversion and limited support (monozygotic twins have higher concordance rates than dizygotic ones) for psychopathy.

A genetic predisposition for criminality and antisocial behavior, which are forms of social risk taking, remains challenged and inconclusive, at best. Some relationship between arousal seeking and engaging in social risk taking may be assumed, which could then lead to labelling those individuals being caught during the act. In the light of such inconclusive findings it is very dangerous to ascribe the term genetic disposition to criminal behavior. It could lead to dealing with criminality as a form of genetic, individual, and unchangeable problem, rather than looking for solutions to societal problems that contribute to criminality.

In relation to risk taking it could be argued, however, that regardless of whether the cause for a link between personality-patterns and criminal activities is genetic or labelling, a relationship exists and thus justifies the assumption of a personality factor in taking social risks. One of the consequences of Eysenck's biological assumptions about different arousability in extroverts than in introverts is that extraverts should prefer situations and sensations which present them with higher levels of arousal, more challenges and more novel experiences.

3.4 Risk Taking and Sensation Seeking

Every situation of perceived riskiness is accompanied by arousal. When looking for unusual, exciting or stimulating experiences, which are by definition "new, or unusual", we are willing to expose ourselves to a higher degree of uncertainty, or risk, than when we engage in familiar, or "boring" activities. In contrast to assumptions of risk being perceived as an undesired event that is to be reduced, or merely more easily tolerated by a stronger physiological system, most people enjoy one form of risk or another. Amusement parks, sports, watching thrillers, gambling, are all activities that are actively searched out and can be intensely enjoyed.

I also state that an exciting, or thrilling sport event, movie, or discussion offers more attraction than a boring one does. The success of exciting events in attracting the interests of the majority of people are sufficient indication, that people do not go there with the hope of not being aroused, or the willingness to endure an increase in undesired arousal to be afterwards rewarded by a drive reduction. It appears obvious that people look for sensations, arousals and risk-experience that provide them with pleasurable feelings. That different people look for different kinds of stimulation, as well as different degrees of arousal, led to the combination of inverted-U arousal theories and personality theories.

3.4.1 Zuckerman's Optimal Level of Arousal Theory

To understand the finding that people actively seek out risky situations and based on sensory deprivation studies, Zuckerman (1979, p.10) described the personality trait of ***"sensation seeking"*** as:

"the need for varied, novel, and complex sensations and experiences and the willingness to take physical and social risks for the sake of such experiences."

The biological components of the theory, the psychological assessment, its phenomenological correlates, the relation to other arousal-personality theories and criticism of the theory will be discussed.

In contrast to Eysenck (1967), Zuckerman (1980) focussed on the biological components of sensation seeking. According to Zuckerman (1983), biological disposition, reactivity and presence of biological correlates of sensation seeking are genetically determined. Fulker et al. (1980) studied 422 pairs of twins and reported a heritability factor of 58% for sensation seeking. Zuckerman (1980; 1991) further listed several breeding experiments and open field observations on mice and monkeys to support his argument that sensation seeking is a genetically determined construct. Observations and physiological studies provide support for the concept as a whole, as well as for the assumed biological correlates of sensation seeking (Zuckerman, 1984; 1991).

As biological correlates of sensation seeking Zuckerman (1980; 1983; 1984) suggests the limbic system, the level of monoamino-oxidase (MAO), gonadal hormones (testosterone), reactivity towards evoked potentials, and the orienting reflex. Stein (1983) argued that catecholamine releas is the basis for intracranial self-stimulation, as described by Olds and Fobes (1981).

Gray et al. (1983) did not agree with this proposition, but argued instead for a purely dopaminerg mediation of the reward system in the brain. Zuckerman (1984) and Stein (1983) argued for the existence of two reward systems, one in the limbic system which reinforces increases in arousal through the release of catecholamines, and another system that rewards decreases in arousal through the release of endogenous opiates (endorphines). Zuckerman (1984) hypothesized that such intrinsic reward systems reinforce both the search for novel stimuli and their mastering.

Zuckerman (1983) described the possible involvement of the level of MAO, being an enzyme that controls levels of monoamines in neurons. MAO is believed to be related to active behavior; MAO blockers can increase activities in rodents (Murphy & Weiss, 1972), as well as in depressive patients (Zuckerman, 1979). MAO levels correlate negatively with sensation seeking (Murphy et al., 1974). Murphy et al. (1977) showed MAO levels to be under genetic influence and that MAO levels of females are lower than those of males. Females and older people are also consistently found to have lower levels of sensation seeking (Zuckerman, 1979).

Smith et al. (1989) showed that high risk seekers showed stronger psychophysiological arousal in response to sexual and violent stimuli than low sensation seekers. Ball and Zuckerman (1992) found that high sensation seekers have better focussed attention in dichotic listening tasks and on shadowing with light tasks. This difference was strongest with the introduction of a secondary novel stimulus, as low sensation seekers' performance deteriorated while high sensation seekers' performance did not.

Zuckerman (1984) further argued that novel stimuli are not rewarding if they differ extremely from previous experiences, but they provoke fear instead. Although Zuckerman (1984) admitted that the environment plays a role in the development of a sensation seeking trait, he sees the dominant influence in genetic and physiological determination (Zuckerman, 1989).

The exact role of nature versus nurture, however, cannot be determined (Martin, 1985; Kafry, 1982). Kafry (1982) studied 69 children from kindergarten to fourth grade and reported that sensation seeking scores were related to observed behavior of a sensation seeking type. He was, however, unable to rule out either model-learning or other early socialization experiences relative to sensation seeking.

Farley (1986) suggested that this personality trait may be determined by genetic make-up, experiences around time of birth, and early nutrition. These findings suggest that sensation seeking may develop already early in life, influences attention processes, but can also be influenced by external stimuli.

3.4.2 The Sensation Seeking Scale (SSS)

Zuckerman, Kolin, Price and Zoob (1964) designed a personality questionnaire to operationalize and quantify the concept of optimal stimulation and arousal (Zuckerman, 1980). The so called "sensation seeking scale" is the result of factor-analyzing a large number of items expressing high or low levels of optimal stimulation or arousal in human preferences, interests and activities. The scale underwent several revisions, and its sixth form has been validated (Zuckerman, 1984b).

Scale V consists of 40 items, relating to one general, second order, sensation seeking factor, consisting of four sub-factors. The factors are:

1) *Thrill and Adventure Seeking (TAS): the seeking of adventure through risky but exciting sports and other activities, such as fast driving;*

2) *Experience Seeking (ES): seeking sensation through the mind and the senses and through a non-conforming life-style;*

3) *Disinhibition (Dis): the seeking of sensation through social stimulation and restlessness when exposed to such situations;*

4) *Boredom Susceptibility (BS): an aversion to monotonous, invariant situations and restlessness when exposed to such situations.*
   (cited from: Zuckerman, 1984, p.414).

The factor structure of Zuckerman's (1971) American sample, was confirmed by Zuckerman, Eysenck and Eysenck's (1978) British sample, and found support in Ball et al.'s (1984) Australian study, Rogers's (1987) study in Canada, as well as Birenbaum and Montag's (1987) Israeli sample. Zuckerman (1984) stressed the high reliability (3 week retest reliability= .94) and validity of the test. A confirmatory factor analysis by Rowland and Franken (1986) produced a similar factor structure on a Canadian sample. These researchers, however, pointed out the "outdatedness" of some items.

In addition to the reported factor structure, Zuckerman et al. (1978), Rowland and Franken (1986), and Ball et al. (1984) reported similar findings in regard to sex and age differences, with men and younger people (age 17-21) scoring higher on the scales. Zuckerman (1979) also found racial differences, with blacks scoring lower than whites on TAS. He explained this finding with the fact that this scale refers to activities like mountain climbing and SCUBA diving, which involve money and are more familiar to white middle-class Americans.

The Sensation Seeking Trait and Phenomenal Correlates

To test whether the concept of sensation seeking holds up in a natural environment and can predict behavior, numerous studies have used the scale and correlated it with various types of risk taking behavior. For example, Rowland et al. (1988) reported that subjective assessments of sensation seeking personalities were done very quickly and corresponded accurately to questionnaire measures, for both familiar and unfamiliar others. In studies on special risk taking groups, the relation between activities and some of the SSS sub-scales have been shown for the use of alcohol (Schwarz, Burkhart & Green, 1978), illegal drugs (reviewed by Kohn & Coulas, 1985), the variety and frequency of sexual experiences (Zuckerman, 1983), food preferences and driving habits (Zuckerman & Neeb, 1980).

Sensation seekers tend to smoke more, believe less in conventional religion (Zuckerman & Neeb, 1980) and volunteer more often for unusual experiences and activities, such as sensory deprivation and hypnosis (Zuckerman et al., 1967), drug research, alpha control, meditation training and gambling classes (Zuckerman, 1979), volunteer for unspecified research, or "unusual" types of psychological experiments (Kohn et al., 1982).

Furthermore, high SSS-score were associated with engagement in risky sports, with higher accident rates amongst skiers and with extended diving times (Zuckerman, 1983b). Correlations of high SSS-scores with increased preference for and participation in high-risk sports were found for activities such as: parachuting (Hymbaugh & Garrett, 1974), hang-gliding and auto-racing (Straub, 1982), motorcycling (Brown et al., 1974), SCUBA-diving (Heyman and Rose, 1980), mountain climbing (Robinson, 1985; Fowler, Von Knorring & Oreland, 1980) and downhill skiing (Bouter et al., 1988). Similar findings were reported (Zuckerman, 1983b) for physical contact sports, by Zuckerman (1979) for "streaking" (running on campus and downtown without clothes on), and by Rowland, Franken and Harrison (1986) for participating in a greater variety of sports. Bouter et al. (1988) reported higher SSS scores for downhill skiers as compared to a control group, but they reported a lower likelihood of accidents for the sensation seekers than for others. This finding was attributed to the finding that the persons scoring higher on the SSS scale were also better skilled.

In contrast, Smith, Ptacek and Smoll (1992) found that being high in sensation seeking did not lead to either fewer or more injuries in school sports, but served as a psychological coping mechanism in reducing the negative emotional and cognitive consequences of sport injuries. The authors further reported that being a high sensation seeker did not lead to more incidences of stress-producing situations in sports, but lowered the troublesome and depressive thoughts after injuries. Low sensation seekers are assumed to attribute their feelings of anxiety to their inability to cope with the situation which then increases their emotional disturbances.

Zuckerman (1979) further found correlations between SSS scores and interest in travel and mobility. Sarason (1978) reported that sensation seeking acted as a moderator variable between negative life changes and subjects distress, with lower sensation seekers experiencing more stress for similar life changes. Similar findings were provided by Cohen (1982). Low sensation seekers were also found to prefer more tranquilizing music (Zuckerman, 1979), and believed more in ESP and psychical phenomena (ESP= Extra-Sensory-Perception) (Davis et al., 1973). Levin and Schalmo (1974) and Looft (1971) reported positive correlations between SSS scores and self-rated liberalism in religious and political terms. Kish and Donnenwerth (1972) reported strong negative correlations (-.82) between sensation seeking and authoritarianism. Newcomb and Mcgee (1991) found that high sensation seekers had more contact with drugs and so called "general deviances". Their longitudinal assessment did allow to perform a LISREL analysis which provided support for a model that revealed the four dimensions of the SSS V as latent structures.

Sensation Seeking and Choices in Life

Consistent with Zuckerman's concept of sensation seeking, as well as with the evolutionary model described above, it is to be expected, that risk seeking or sensation seeking subjects choose jobs involving high levels of change, flexibility, and jobs involving some kind of risk (Zuckerman, 1979). A study of individuals in risky professions (e.g., firemen, mountain-rescue squad men), risky sports (race car drivers, mountain climbers), and a non-risk control group (Zaleski, 1984) showed that the sport-risk group scored higher on the SSS than the professional-risk group, which scored higher than the non-risk group.

Musolino and Hershenson (1977) showed that subjects with occupations that placed themselves or others in jeopardy (e.g., air-traffic controllers, pilots) scored significantly higher on the SSS than civil servants and college students. Kish and Donnenwerth (1969; 1972) reported negative correlations between SSS scores and clerical interests, and positive ones with scientific interests. Furthermore, they found significant positive correlations between sensation seeking and perceived likelihood of being members of professions related to helping and social interactions

Best and Kilpatrick (1977) found that rape crisis counsellors scored significantly higher on the SSS than paediatric nurses. Furthermore, air line pilots and air traffic controllers scored significantly higher on the SSS than college students or civil servants (Zuckerman, 1979). Farley (1986) reported that people high in thrill-seeking were also sometimes outstandingly creative, destructive and even tended to criminal behavior. Franken and Rowland (1990) found that high sensation seekers had significantly more flamboyant, sensual and varied fantasies than low sensation seekers.

With criminal behavior regarded as one form of risk taking, it is not surprising that Zuckerman (1979), and Emmons and Webb (1974) found the DIS (disinhibition) scores of the SSS correlating with psychopathological behavior. Blackburn (1969; 1979) reported correlations between psychopaths scoring highly on a sociability scale with Thrill and Adventure Seeking (TAS) and DIS. These correlations were higher than for non-psychopathic criminals and for psychopaths scoring low on the sociability scale. Perez and Torrubia (1985) reported significant correlations between the SSS subscales and antisocial behavior. Farley and Farley (1972) reported more escape attempts and disobedience by female prisoners scoring highly on the SSS. Zuckerman (1979), however, reviewed several studies showing no significant relation between either delinquency, or criminality and SSS, and pointed out the problems related to research linking personality and criminal risk taking, discussed before. The relations described, may, thus, not be generalizable to other groups.

Sensation Seeking and Other Personality Theories

Eysenck and Eysenck (1976) argued that their impulisivity dimension contains the sensation seeking trait. The assumption led them to construct an impulsivity questionnaire containing a risk scale with items similar to the TAS scale of the SSS, and correlating them (r=.57). The Neuroticism scale, however, was not related to the SSS (Eysenck & Zuckerman, 1978).

Jackson's (1967) Personality Research Form (PRF) includes Harm Avoidance, Change and Impulsivity scales which somewhat overlap with SSS-subscales. Items in the Harm Avoidance scale are similar to TAS items, except that they are worded to indicate avoidance of risk, rather than sensation seeking. Positive correlations between SSS and Aggression, Dominance, Exhibitionism, and Play were found, and a negative correlation with the factor "*Cognitive Structure*", all being subscales of Jackson's PRF. A person scoring highly on cognitive structure is described as:

- *"Does not like ambiguity and uncertainty in information; wants all questions answered completely; desires to make decisions based upon definite knowledge rather than guesses or probabilities."*

- *"Play" is described (p.7) as: "Does many things just for fun; spends a good deal of time participating in games, sports, social activities and other amusements; enjoys jokes and funny stories; maintains a light-hearted and easy-going attitude toward life"* (Jackson, 1967, p.6).

Therefore, both the negative correlation of SSS with cognitive structure, as well as the positive one with Play, fit the description of a sensation seeker or risk-taker, although Jackson et al.'s (1972) four dimensions of risk taking differ somewhat from Zuckerman's.

Zuckerman (1979) found correlations in the expected direction with Cattell's (1957) 16PF test, with the California Personality Inventory (Gough, 1957), and the Minnesota Multiphasic Personality Inventory (MMPI) by Hathaway and McKinley (1951). Neary and Zuckerman (1976) reported positive correlations of sensation seeking with the "strong" nervous system type by Nebylitsyn (1972). Furnham (1984) and Ortega and Pipal (1984) found correlations of the Sensation Seeking trait and challenge seeking with "Type A" measures. Burkhart, Schwarz and Green (1978) reported negative correlations between SSS scores and anxiety dimensions.

Kohn, Hunt and Hoffman (1982) and Goldman, Kohn and Hunt (1983) studies showed significant correlations of high SSS scores with the reducing component of Vando's (1974) Augmenter-Reducer Scale (Augmenter=Arousal Seeker, Reducer=Arousal Reducer). They argued that the concept of Augmenting-Reducing gives different results and is used differently, dependent upon the type of operationalization. A solution to this alleged contradiction was suggested by Goldman et al. (1983). It lies in the relation of both concepts to the "strength of the nervous system" approach, which indicates that sensation seekers are less arousable and thus augment their physiological experience, while the functional endurance aspect leads the "strong" nervous system types to reduce arousal.

The relationship between these theories, remains, therefore, unresolved, as Goldman et al. (1983) findings were partly contradictory to those found by Zuckerman (1979). Dragutinovich (1987), as well as Lolas et al. (1989), reported significant correlations between the SSS and the reducing component of Vando's (1974) Augmenter-Reducer Scale, thereby further arguing against Zuckerman's findings of SSS correlating with arousal augmenting.

Criticism of Zuckerman's Theory of Sensation Seeking

The strongest criticism of Zuckerman's concept lies in the correlational nature of the reported studies. Although the studies finding plausible correlations between risk taking and biological factors are numerous and somewhat consistent, conclusive experimental assessments still need to be done. Also, Paisey and Mangan (1980) reported a factor structure different from Eysenck's three personality dimensions, and that Zuckerman's SSS V factors were not replicated in their confirmatory factor analysis at all.

Baldwin (1984) raised objections against the heritability of sensation seeking, since all reported findings can also be explained by social learning theory. Furthermore, the size of environmental influence cannot be estimated correctly by heritability coefficients. Barratt (1984), and Barratt and Patton (1983), argued that people can be anxious and sensation seekers at the same time, which casts doubt on the neurotransmitter status within the brain. Furthermore, Barratt (1984) stated that MAO influences many centers in the brain that are presumably involved in various activities and dimensions of personality, thereby leaving open, whether MAO influences sensation seeking directly or via some other process.

Clark (1984) argued that inter-species comparisons may not be adequate because of obvious limitations in assessing and comparing animals sensation seeking attitudes and behaviors, and because of differences between open field observations and laboratory experiments. Clavier (1984) doubted monoamines to be the neurotransmitters rewarding sensation seeking, since brain centers important for self-rewarding are low in catecholamines.

Izard (1984) criticized the lack of emotional aspects in the SSS, as emotions seem to play a major role in risk taking. Katz (1984) argued that if sensation seeking is a biological necessity it would be performed by every organism under certain conditions. The concept of a trait is therefore inadequate. Wohlwill (1984) argued against the adequacy of some items, and Zuckerman (1984b) admitted the problem of socially desirable answers to the items in scales I to V, one reason why he designed form VI.

Andresen (1986; 1990) translated form V of the scale into German but did report poor psychometric results on 223 subjects. He suggested to use other forms of scales, such as the Multidimensional Inventory of Situational and Actional Preferences (MISAP II) in non-Anglo-American cultures.

Stacy, Newcomb and Bentler (1991) followed 584 American adolescents in a 9-year longitudinal study and found that earlier sensation seeking, operationalized through SSS V, correlated strongly with later sensation seeking. However, a structural equation modelling approach revealed, that socio-psychological factors, operationalized through scales for social conformity, peer deviance, social support, emotional distress, drug abuse and former sensation seeking reports, could also explain the later found sensation seeking score. The authors conclude that sensation seeking seems to have a developmental aspect in addition to genetic components.

Overall, Zuckerman's approach of combining both physiological and psychological concepts and assessing data, both via personality questionnaires and physiological measurements has received positive reviews (Eysenck, 1984; Strelau, 1984). In relation to arousal theories and risk taking, the concept offers up to now the most elaborate personality component that finds support by biological data. Thus, risk taking behavior seems to have a personality component that can be measured by its biological correlates, as well as by self-report.

3.5 Reversal Theory

After having discussed the relationship of arousal and risk taking in single-peaked, optimal level of arousal personality theories, some context dependent risk taking behavior patterns still remain unexplained. Different situational frames and expectations (e.g., sudden emotional shifts), as well as different personalities may very well interact to produce the sometimes opposite behaviors showed by the same person in objectively similar situations. For example, whether a person is expecting to be either promoted or fired may alter the willingness to raise a concern about identical poor working hours or pay. Reversal theory attempts to explain such phenomena. I shall present the concept first in examples, next theoretically, discuss its relation to other risk-related research, and raise some critical remarks.

3.5.1 Pleasant and Unpleasant Risk Taking (Reversal Theory)

Imagine two drivers speeding on a motorway. One of them could be speeding because she has to reach a certain destination in time and she is late. To be there on time is important to her, because her job depends on it. Thus, she is extremely anxious about being late. Her speeding behavior is planned in order to achieve a certain goal and she is continuously evaluating her chances of achieving the goal. When she is ahead of her schedule and reaches her destination on time, her anxiety will diminish and she will experience a pleasant feeling of relaxation after a period of high tension.

Now consider the other driver. He may be driving his car fast simply because he enjoys doing so. He has no destination to reach in time and he is simply driving " for the fun of it", because he is exhilarated by the speed, arousal and risk, or because he enjoys testing his skills and feeling that he is master of the situation. When he reaches a place where he has to slow down, he will most likely experience it as not pleasantly relaxing, but as an unwanted interruption of his preferred activity, and if the obstruction continues, he will most likely get bored, or agitated and engage in a compensating behavior to regain the pleasant excitement. The second driver is not motivated by any special goal, except the goal of experiencing something pleasant there and then. Although both drivers may show the same behavior, the emotional experience and the meanings associated with the behavior are totally different (examples from Apter, 1982, p.6).

Now consider two people gambling. One may be doing so, because he sees it as the only way to get the money he urgently and quickly needs. A close game, endangering his win and future, will be experienced as unpleasant and anxiety arousing. The other player, however, plays just "for fun" and she enjoys the risky situation as a pleasantly exciting one.

The Concepts of Telic/Paratelic Dominance

To apply the terminology of reversal theory, the first actor was goal-oriented, or ***Telic*** (after the Greek word 'telos'= goal or end), while the second actor was "activity-oriented", or ***Paratelic*** (after the Greek words 'para' = alongside and 'telos').

Apter (1982, p.47) stated:

"The telic state is defined as a phenomenological state in which the individual is primarily oriented or feels the need to be primarily oriented toward some essential goal or goals. The paratelic state, in contrast, is defined as a state in which the individual is primarily oriented or feels the need to be primarily oriented toward some aspect of his continuing behavior and its related sensations."

Depending on the state one is in, he/she will engage in homeostatic action around either one of the target experiences. The concept of "bistability" (rather then single-peaked homeostasis) (Apter, 1982), contrasts with other optimal level of arousal and homeostasis theories by introducing a second optimal level of arousal, to which the person switches, depending on the situation. This bistability notion supposes that two dimensions of arousal-motivation exist.

One of these dimensions is the goal-oriented (telic) mode ranging from relaxation as a pleasant state with low arousal levels, to anxiety, experienced at high arousal levels. The other dimension, the "activity oriented" (paratelic) mode ranges from boredom at the low arousal end, to excitement at high arousal levels. The experience of given arousal levels within the two modes is directly reversed (pleasant/unpleasant). As many factors can quickly switch one mode to another (telic/paratelic), Smith and Apter (1975) named the concept *The Theory of Psychological Reversals*.

To illustrate the concept of reversal, I want to refer back to the two drivers. The driver who was speeding for the fun of it may very well encounter an oncoming vehicle while overtaking, or a police car at the side of the street. Both situations will increase the arousal level, but will most likely be experienced as anxiety provoking rather than as pleasantly exciting, for the driver has most likely switched into the telic mode. The goal will be to avoid a collision or a ticket for speeding.

The other driver, however, who was speeding to be at an appointment in time, may realize that she is ahead of schedule and switch a the paratelic mode, where she changes her interpretation of her arousal due to speeding from anxiety provoking to pleasantly exciting. Both drivers, however, will adjust their behavior such that they experience their preferred arousal level corresponding to the telic/paratelic modes.

The telic state is further split into sub goals, such as: physiologically essential, socially essential, self-esteem essential, well-being of the family essential. These goals are imposed, or unavoidable (Apter, 1982, p.50) in contrast to freely chosen activities, such as parachuting, racing, gambling or motorcycling. In the telic state the individual supposedly feels to be reactive, in the paratelic state proactive. This also means, that in the telic state the person wants to finish the activity as quickly as possible to accomplish the goal (e.g., exams), whereas in the paratelic state the individual does not want to reach a certain goal and wants to stay in the situation as long as possible (e.g., sexual activity).

Reversal theory states that every person is telic as well as paratelic in different situations, and that the experience of pleasant excitement can be instantly reversed into unpleasant anxiety. Possible causes of the reversal process have to be specified. Biological needs, for example getting hungry or thirsty, may interrupt the paratelic activity by forcing a goal upon the person. Similar reversals into the telic mode are brought about through extremely dangerous events with excessively high levels of arousal, or threats to the self or one's family (Apter, 1982). At the same time, social cues such as laughter, a football field or a theatre are likely to induce a paratelic mode, while a scream or a police station or the dentist are likely to produce a telic mode (Apter, 1982). The paratelic mode is therefore only accessible when the telic mode does not require immediate reaction.

Apart from these events, frustration and satiation can also induce a reversal of the experienced state. A frustrated or satiated telic person will experience boredom and look for higher arousal levels, while a paratelic person, who has had enough arousal may switch into a telic mode and look for pleasantly relaxing goals. Apter (1984) states, that the process of satiation reversal is an innate one, which builds up in strength until a reversal is brought about. Apter (1984) calls it an "innate dynamic for change".

Reversal mechanisms are used to explain phenomena such as sexual dysfunction in spite of high arousal. Apter (1982) argued that the predominantly paratelic activity of sex (unless done for the sole purpose of producing offspring) may become a telic one. If high expectations on having to produce an orgasm, or to perform to some - real or imaginary - standards occur during intercourse, the experienced arousal will no longer be pleasantly exciting, but rather anxiety provoking and unpleasant. The higher the arousal the more unpleasant the experience after a reversal and the worse the performance. This will then perpetuate the sexual dysfunction, as the person feels inadequate both in performance and feelings.

While it is assumed, that excessively high levels of arousal will automatically cause a reversal even in paratelic dominant persons, there are supposedly individual differences (Apter, 1982). The levels at which switches occur, the strength of emotional experiences, the speed with which satiation builds up, the skill, and the attributional schemata may all explain individual differences in the more telic or paratelic experience of arousal (Apter, 1982; 1984). Furthermore, the telic/paratelic states themselves may be differently developed in different people. The thresholds, slopes, peaks and form of the experienced arousal curve, not only differ between subjects, but also within subjects across situations (Apter, 1982).

To assess whether a person tends to be more often telic than paratelic, and vice versa, Murgatroyd, Rushton, Apter and Ray (1978) developed the "Telic Dominance Scale" (TDS). This 42 item questionnaire yields a telic/paratelic score. The scale of 42 items has been broken down by experimenter judgement and factor analysis with varimax rotation into three subscales of 14 items each, named *Seriousmindedness, Planning Orientation,* and *Arousal Avoidance.* The rationale behind this particular split is given by Murgatroyd et al. (1978) as being the best split to differentiate between predominantly telic and paratelic personalities.

Apter (1984, p.278) defined the subscales of the questionnaire as:

> ***Seriousmindedness*** *is the frequency with which an individual tends to be oriented toward goals which he or she sees as being important, rather than merely acting as the endpoints for ongoing activities enjoyed in themselves.*
>
> ***Planning orientation*** *is the frequency with which an individual tends to plan ahead and organize him or herself in the pursuit of goals rather than taking things as they come.*
>
> ***Arousal avoidance*** *is the frequency with which an individual tends to avoid situations which generate high arousal rather than avoiding situations in which arousal is likely to be low".*

Apter (1982) pointed out, that although the reliability coefficient was 0.68 for their entire sample, any individual can be telic on one subscale and paratelic on another. Shifts in "Dominance" of scales have also been reported, with subjects being telic on one scale and paratelic on another.

Research on Reversal Theory

Apter (1982) applied reversal theory to about 170 studies from social psychology, psycho-physiology, animal learning, cognitive psychology, psycho-linguistics, perception, comparative psychology and psychometrics. Findings of these studies could be generally explained by reversal theory, the nature of the explanation remained, however, a posteriori. Reversal theory has also been used to explain phenomena, such as aesthetic experience (Apter 1982; 1984), sexual dysfunction and perversion (Apter and Smith, 1979), antisocial behavior in childhood, adolescence, and old age (Apter, 1982), the generation of religious myths and rituals (Apter, 1982) and enjoying humour as a special form of paratelic arousal (Apter, 1982). Research has also been designed especially to test the validity of reversal theory and its predictions.

Apter (1976) let 67 subjects rate 50 everyday situations both on experienced arousal, as well as on hedonic tone. He found that of the eleven highly pleasurable situations, four loaded highly on arousal, three average and four loaded low on arousal. This result was incompatible with optimal level of arousal theories, which would have predicted an inverted-U shaped function, but it was predicted by reversal theory. Wicker, Thorelli, Barron, and Willis (1981) found support for a relation between humour and the combination of high arousal and paratelic state.

Svebak, Storfjell, and Dalen (1982) found that electromyographic activity in the task-irrelevant musculature of telic subjects increased, indicating a general tension, while it remained constant in paratelic subjects, pointing to a relaxed status. Under conditions of threat, telic subjects showed a significantly higher increase of heart rate than paratelic dominant subjects. The authors also conducted interviews which revealed playful, spontaneous trends in paratelic and mundane tendencies in telic dominant subjects. Reversal theory has also been utilized in clinical psychology with biofeedback and relaxation training (Svebak & Stoyva, 1980).

Martin, Kuiper, Olinger and Dobbin (1987) examined the relationship between stress and telic/paratelic dominance. A multiple regression analysis revealed that telic subjects felt more stressed if recent stressors where not resolved, while paratelic subjects felt less stress. This result was supported by corresponding levels of salivary cortisol, a biochemical substrate related to stress (Baum, Gruneberg & Singer, 1982; Frankenhäuser, 1986).

Furthermore, telic subjects enjoyed a video game more if nothing was at stake, while paratelic subjects showed no difference. Finally, telic subjects showed a linear increase in mood disturbance with increasing negative live events and/or daily hassles. Paratelic subjects, however, showed a curvilinear relationship, reporting less mood disturbance at moderate levels of hassles and negative life events and higher mood disturbances at low levels of negative life events and hassles. The authors concluded that paratelic subjects seem to thrive on moderate levels of stress experiencing the greatest enjoyment and challenge when faced with moderate levels of stress and problems, while getting bored at easy tasks.

Reversal Theory and Other Personality Theories

Apter (1982) compared the paratelic dominant person with the "hardy" person in Kobasa's (1979) concept of hardiness, where hardy persons perceive problems more as challenges than threats. Similar comparisons to Dienstbier's (1989) "toughness" can also be drawn, with paratelic dominant people being more toughened through playful experience, while telic persons avoid additional or unusual sources of arousal. Apart from similarities to hardiness, significant correlations were reported between the telic/paratelic concept and other measures of personality.

For example, Cowles and Davis (1985), and Murgatroyd (1985) correlated the "Arousal Avoidance and the "Planning Orientation" subscales of the TDS with the Sensation Seeking Scale Form IV (Zuckerman, 1974). Studies revealed correlations of -0.40 between both the "Seriousmindedness" and the "Planning Orientation" subscales, the Eysenck Personality Inventory (Eysenck & Eysenck, 1976) and a correlation of -0.38 with the Reducer-Augmenter Scale (Vando, 1974). Martin et al. (1987) pointed out similarities between reversal theory and *"Strength of the Nervous System Theory"* (Nebylitsyn, 1972; Strelau, 1982; 1983).

Murgatroyd et al. (1978) reported further correlations between their scale and the trait measure of anxiety by Spielberger, Gorsuch and Lushene (1970), as well as Robinson's (1961) "Need for Achievement" measure. Thus, paratelic subjects seem to be sensation seekers, extraverted, arousal reducers, and have a high hope for success.

Apter (1984), however, reported no correlations with the EPI, and no significant correlations with the Type A/B measure by Jenkins, Rosenman and Friedman (1967). Contradictory results, as well as moderate correlations are interpreted in various ways. Martin et al. (1987) suggested that the moderate correlations are either an indication for the unique contribution of reversal theory to this area, or that the reversals have taken place and thereby changed the results (Apter, 1982).

Apter (1982, 1984) stressed the distinctiveness of reversal Theory from other arousal and homeostasis theories. He utilized Deci's (1975) review of "intrinsic/extrinsic motivation" to argue that telic/paratelic dominance is independent of intrinsic motivations of the central nervous system. Furthermore, Apter (1982) distinguished reversal theory from Kruglanski's (1975) concept of endogenous/exogenous attributions with the argument that attributions are different from goals and that Kruglanski's association of arousal with positive and negative experiences is contrary to findings in reversal theory (Apter, 1982).

Some problems with reversal theory have to be noted. The assumption of a telic/paratelic reversal in certain situations, depending upon the individual's interpretation and general personality type, makes it hard to disprove the theory. It seems to be impossible to differentiate whether a reversal has occurred, or whether the person is of either the telic, or paratelic type. Can multiple reversals in the same situation be accounted for? How do defense mechanisms influence the reversal process? The difficulty with the questions just raised is somewhat supported by the scarcity of studies trying to induce reversal processes under controlled conditions, while keeping the trait-variable constant. In total, a great need exists for experimental testing of the hypotheses raised by reversal theory.

3.5.2 Reversal Theory and Risk Taking

Nevertheless, reversal theory offers an interesting perspective for risk taking behavior. While being in the telic mode subjects will take the risks necessary to reach their goal. The person driving to the job takes the risk of a traffic accident and experiences this as anxiety provoking, as accidents are detrimental to reaching the goal. The person in the paratelic mode, however, will look for risky situations, as the arousal produced through the riskiness will add excitement, which will feel even more pleasant. Even if the person switches shortly into a telic mode to avoid a collision, immediately afterwards, a reversal into the paratelic mode will make the person feel even more "alive" and excited, as arousal levels were extraordinarily high and no harm was done. Since the arousal subsides relatively slowly, excitement may continue to be felt for a fairly prolonged time, thereby providing continuous reinforcement for risk taking.

Furthermore, feelings of mastery, self-esteem, maybe social recognition, or admiration may add to this positive experience, so that for paratelic risk-takers, the net benefit of their utility assessment of risk taking will be positive and thus reinforce the behavior. Apter (1982) further argued that deliberate risk taking behavior is undertaken to increase pleasant arousal by using novelty and uncertainty. Individuals are likely to overcome personal limitations through attempts at novel, or risky activities. A person discovers more about the world through personal or fantasy experience, for example in form of fiction, movies or fantasies. Predominantly paratelic persons would fit more to this description than telic ones.

Furthermore, from an evolutionary perspective, Apter (1982) argued that exposing oneself to risk increases one's ability to cope with novelty and change, a skill that is highly necessary in changing and uncontrollable environments. The concept of a biologically supported paratelic risk taking tendency is further illustrated by the actions of explorers, professional hunters, warriors and pioneers, who must have a high level of paratelic and risk inclination, since their activities reduce their own chance of survival, but increase the chance of survival for their respective communities. As an interesting sidepoint, Apter (1982, p.303) predicted, that with decreasing opportunities for hunting, exploration and wars, the presence of these people in a community may become a source of continual internal threat which will ultimately militate against its successful survival. Present developments in the USA may be indicative of this development.

One advantage of reversal theory is that it is equipped to deal with inconsistencies, both across subjects, as well as within subjects. As reversal theory predicts switches from telic to paratelic states, depending on the perception of the situation, the mood and other extrinsic factors, a person may very well behave differently in the same situation, given a different personal interpretation or state. This distinguishes reversal theory from pure trait concepts, such as Eysenck's extravert personality, who is always an extravert (Apter 1982).

The importance of perceived control, familiarity, security and predictability to induce the paratelic state has also been stressed by Apter (1982). This in turn means, that in case of paratelic arousal, we deal with a kind of planned arousal in a subjectively controlled situation. This assumption is strongly supported by Piët (1987), who reported that stuntmen experienced their profession as generally risky but individually and situationally controlled, and that they spent a lot of their time with making the situation controllable, while maintaining the pleasantness of arousal.

Generally, in sports or other hobbies, freely chosen risks, challenges and competitions are attributed by Apter (1982) to a paratelic state. Whether attempting to have the best front garden, or sail around the world single-handedly, both include some risk, uncertainty, challenge and increased arousal. This arousal may be experienced as pleasant, when in a paratelic mode and as unpleasant and frustrating when a reversal to the telic state followed a lack of success. Reversal theory interprets risk taking both as a trait, with people being predominantly telic or paratelic, and as a state, because people can switch from one mode into another, if the situation so requires. As, however, most studies on reversal theory and risk taking are a posteriori explanations or correlational between subject studies, Martin et al. (1987, p.981) stated:

"Thus, a more thorough analysis will require following subjects over time to examine changes in metamotivational states relative to stress levels and to observe attendant changes in mood, performance, and physiological reactivity."

After having predominantly discussed personality aspects of risk taking, I conclude that there is a substantial amount of evidence that an inherent, genetically supported and evolutionarily favourable component of risk taking exists. This propensity to take risks is physiologically and psychologically rewarded just as the arousal reduction mechanisms in cases of arousal as a result of risk taking beyond the individual's perceived optimal level of arousal. The degree to which personality aspects do determine or explain behavior is, however, undetermined and varies extremely given the different experimental conditions. In addition, as an individual's assessment of costs and benefits of risk taking in any given situation, experimental or real, does determine behavior at least as strongly as a personality disposition may do, I shall now turn to situation oriented approaches and rational decision making theories of risk taking.

Chapter 4: Which Situational Factors Influence Risk Taking Behavior ?

After having discussed biologically based risk taking theories, I want to turn to cognitive theories where risk taking is discussed as decision making under uncertainty. These theories assume that people make rational choices to maximize their personal profit, while tolerating risk as an unwanted by-product of uncertainty.

4.1 Decision Making Under Uncertainty

The area of ***rational decision making*** is described very comprehensively and is supported by sophisticated research. Neumann and Pollitser (1992) for example, discussed advantages and limitations of such theories from a decision-researcher's point of view. They discussed problems arising from people not adhering in their judgement to the underlying necessary mathematical assumptions in utility theories, such as the reducibility and independence axioms. The decision logic of the average person in day to day life clearly differs from that predicted by optimal decision making theories. Part of that shortcoming may be due to the fact that physiological and emotional components, as well as individual differences remain largely undeveloped in such models and are even hardly discussed in any review of utility theories.

In the following, I shall shortly describe the historical development of ***Expectancy x Value*** theories, their major proponents and criticisms. Some recent suggestions to include emotion and psychophysiological components and the need to combine cost/benefit or expectancy x value theories with other risk taking theories will be stressed.

4.2 Historical Development of Expectancy x Value Theories

Many attempts have been made to express risk formally using mathematical models. The first explicit mathematical theory of risk was developed by Daniel Bernoulli (1738/1967). He proposed that risk is evaluated in terms of the mathematical expectations of the subjective values of the various possible outcomes - in other, modern words - the expected utility (Lopes, 1984). Bernoulli (1738/1967, p.25) described utility as a psychological construct describing the common intuition that:

"any increase in wealth, no matter how insignificant, will always result in an increase in utility which is inversely proportionate to the quantity of goods already possessed".

Thus, the receipt of $100 means more to a poor person than to a rich one. This also means that for Bernoulli the expected utility of a gamble would always be less than the utility of a sure sum, equal to the gamble's expected value (Lopes, 1984, p.482.) The concept of subjective utility and expectations proved difficult to measure, and therefore Bernoulli's theory was not widely used in its original form until very recently. Other measures of risk prevailed (Lopes, 1984). For example, as early as 1786, Tetens proposed to use one-half of the mean deviation of pensions as a possibility to quantify risk in the payment of retirement money.

Libby and Fishburn (1977) summarized many of the suggested measures of risk in business decisions and they named for example: mean-variance trade-off, mean-semivariance trade-off, mean-standard deviation trade-off, mean-probability of loss trade-off, probability of "ruin", etc.. In Bayesian terms, risk was expressed as the expected value of loss (Schäfer, 1978).

Mathematical utility theories took a step forward with the work of Von Neumann and Morgenstern (1947). In the second edition of their book "Theory of Games and Economic Behavior" they picked up on Bernoulli's theory, but expressed and measured utility in a very limited way, compared to Bernoulli's 1738 original suggestion. For them, utility was measured and defined as money under risk. Thus, risk attitude was transformed to a measure of people's subjective value of money (Lopes, 1984).

Subjective expectations were harder to measure, and even considered "mystical" or "nonsensical" by some (Savage, 1972; Lopes, 1984). It is not surprising, therefore, that the theories building upon the work of Von Neumann and Morgenstern (1947) used monetary utilities and rational choices in laboratory situations to advance their concepts of risk taking under conditions of uncertainty. They preferred to use the term uncertainty to reduce the influence of emotional factors in the commonly less specifically used term "risk" (see also Cooper, 1987).

At the same time, work in the areas of decision making, expectations as well as motivation, for example work by Lewin (1938), Hull (1943), Miller (1948), Spence, (1956) Tolman, (1955) and Festinger (1957) sparked the development of Atkinson's (1957) theory (Atkinson, 1982). Atkinson (1957) described the relation between achievement motivation, risk taking, expectancy of success and value of success and stated that the performance level should be greatest under high uncertainty (e.g., 50% chance of success). Persons with a high need for achievement should take intermediate risks, while high and low levels of risk are preferred by people with a low need for achievement. People high in need for achievement would not prefer easy tasks, as easy tasks cannot provide feedback about abilities. People high in need for achievement would not choose hard tasks either, as the probability of failure is high. Intermediate levels of risk and difficulty, on the other hand, offer a perfect challenge, and the chance to test their ability. People with low need of achievement (also called fear of failure) would choose easy tasks, to have success, or difficult tasks to have an excuse for failing.

Atkinson's assumptions have found support in studies on school children (McClelland, 1958), and entrepreneurs (Meyer, Walker & Litwin, 1961). Changes to the theory performed in the following 30 years are summarized by Atkinson (1982), while key concepts of expected value and motive strength have persisted.

They inspired the development of *Subjective Expected Utility Theory* (Edwards, 1954; 1961; Vroom, 1964), *Portfolio Theory* (Coombs, 1972), the somewhat related approaches in *Conflict Theory* (Janis & Mann, 1977; Mann & Janis, 1982; Lopes, 1981; 1983), *Prospect Theory* (Kahneman & Tversky, 1973; 1979; Tversky & Kahneman, 1984), and *Affective Balance Theory* (Grossberg, 1980; 1982).

Furthermore, there are suggestions for improvement of existing subjective expected value theories by Bell (1985) who demands to include factors such as *disappointment* or *regret* or Fishburn's (1977) notion of *aspiration levels*. This list is by no means exclusive, as, already in 1981, Fishburn counted 28 different formal versions of expectancy x value approaches, which are also called cost/benefit analyses. The development of these theories will be outlined by describing some of the most prevalent above named approaches.

4.3 Modern Expectancy x Value Theories

4.3.1 Subjective Expected Utility Theory

Subjective Expected Utility theory re-introduced the notion of subjectivity to decision making, but assumed humans to be purely rational decision makers. Subjects were expected to maximize their financial profit under all conditions. For example, the expected utility of a gamble is found by multiplying the subjective probability of each outcome, with the utility of the outcome and summing all outcomes. The decision maker is then assumed to take the outcome with the highest expected financial-statistic utility (Isen, Nygren & Ashby, 1988). This assumption was proven to be wrong, since Kahneman and Tversky (1984) could show that risky decisions under the condition of the same mathematical probability of success depend both on the wording and the emotional frame of the problem given. Also, people's concept of probability, randomness, regression, correlation and causality are not always rational, but often biased and deviant from *objective* estimates or expert ratings (Kahneman & Tversky, 1973; Payne, 1976; Schoemaker, 1982; Slovic, 1967; Slovic, Fischhoff & Lichtenstein, 1977).

Subjective Expected Utility approaches were criticized by Coombs (1972) for a lack of theory and for avoiding psychological reality. Coombs (1975) continued his critique by stating that Edwards (1954), Vroom (1964) and Fishburn (1970) made the notion of risk superfluous since subjects just maximize their utilities (Coombs, 1972). Not only do Subjective Expected Utility theories ignore the reality of emotions in decision making and risk taking, but Subjective Expected Utility theory also expresses a *moral expectation* for the decision makers, namely that they want to maximize the profit. Other approaches to decisions, such as intrinsically motivated ones are not considered and cannot be accounted for (Schäfer, 1978).

4.3.2 Portfolio Theory

Coombs (1972) and Coombs and Huang (1970) proposed instead *portfolio theory* of risk that deals with risk preference. It was assumed that people have in a given situation a preferred, or ideal level of risk for any given expected value. As the risk departs to either side of the single-peakedness it is liked less by the decision maker and the likelihood for choosing either one decision is reduced. The level of risk and the expected value are thus optimized. Aschenbrenner (1978; 1981) showed, however, that single-peaked preferences are not the only decision process, but that other functions are utilized as well.

Aschenbrenner (1978b) reported that portfolio theory could not account for the results in a money gamble game. Furthermore, Björkman (1984), Lopes (1984), and Fischhoff et al. (1982) criticised the theory for assuming that people always make rational (=material) choices and that people are able to assess *objective* probabilities of risk correctly; - an assumption that has often been shown to be inaccurate. Lopes (1983) placed doubt on the whole concept of rationality in relation to risk expectations. She equated people's intuition about risk with intuition about distributional inequality of values and argued for the use of objective distributions, rather than subjective expectations (Lopes, 1984).

4.3.3 Prospect Theory

Prospect theory is the utility theory that made subjective expected utility and portfolio theory's shortcomings very obvious and offered alternative interpretations (Kahneman & Tversky, 1979). The finding that situations involving loss tend to produce risk averse choices, while situations involving gains tend to increase risk inclined choices could not be readily explained by portfolio or subjective expected utility theories (Fischhoff et al., 1982;

Grossberg & Gutowski, 1987; Tversky & Kahneman, 1981). Prospect theory introduced the use of decision-weights to account for the value functions in risky choices. The exponential function describing risk averse decisions on the *loss* side is shaped convexly, while it is shaped concavely when the decisions regard gains (Isen, Nygren & Ashby, 1988).

The finding that phrasing options of the same mathematical probabilities in positive or negative words (e.g., 60% lives saved vs. 40 % death rate) produces opposite decisions, is explained by prospect theory through the use of weights. Potential heavy losses are *overweighted*, therefore people avoid high possible losses (Tversky & Kahnemann, 1986).

Schurr (1987) provided experimental support for Tversky and Kahneman's (1981) theory. Schurr (1987) showed that groups of professional buyers reacted to situations framed as potential losses with risk seeking and to situations framed as potential gains with risk averse behavior. This risk seeking behavior can also be described as minimizing maximal loss, which would not constitute risk seeking as such, but rather a less risk avoiding tendency (see Tversky & Kahneman, 1984; Fischhoff et al., 1982). In general terms, in contrast to subjective expected utility theories, which suggest that people chose their risk taking behavior in relation to potential gains and losses based on absolute amounts, prospect theory suggests to use a reference point, or anchor value (Isen, Nygren & Ashby, 1988).

In spite of numerous supportive studies for prospect theory in game-playing behavior (reviewed by Pruitt & Kimmel, 1977), criticism has been raised, even by Tversky and Kahneman (1984) themselves. One objection mentioned is that prospect theory is an algebraic, static theory that relies on group choice data. Thus, it does not account for individual decisions or information processing that underlies decision making under uncertainty (Grossberg & Gutowski, 1987). The same authors, as well as Neumann and

Politser (1992) listed a number of well confirmed forms of *irrational* behavior that prospect theory cannot account for. Preference reversal, for example, refers to the robust phenomenon (Grossberg & Gutowski, 1987) that subjects judge two objects as being worth $10 and $8 if presented individually, yet prefer the object for $8 when asked to choose between them. This behavior violates the maximization assumption of current utility theories. The need to include non-material gains into utility equations becomes evident.

4.3.4 Conflict Theory

Mann and Janis (1982), introduced conflict theory to address the criticism that other expectancy theories fell short on describing the cognitive processes involved in decision making under stress, on information acquisition and assimilation, and the formulation of new alternatives. Conflict theory (Janis & Mann, 1977) split the decision making process into antecedent conditions, mediating processes and consequences of psychological stress. This model accounts for some of the necessary adaptation processes towards urgent needs. It further specifies coping mechanisms and how they are established and activated, and when stress disturbs or facilitates decision making. Conflict theory introduced a rather well elaborated decision tree process model, yet it remains oriented to material profit only.

4.3.5 Rubikon Model

Heckhausen (1977; 1980) extended expectancy models into a four-stage sequence of events. The anticipated outcome of the event is the pivotal point of Heckhausen's approach (Feather, 1982). Heckhausen (1980) defined expectations as situation-outcome expectancies, action-outcome expectancies action-by-outcome expectancies and outcome-consequence expectancies. The focus on the different type of expectancies, also stressed by

Bandura (1977), further developed utility theories towards an excellent descriptive model of *rational* decision making under conditions of uncertainty. Heckhausen's (1977; 1980) cognitive motivation model explains other related concepts, such as achievement motivation and motivation in general. Although not explicitly stated, the underlying assumption of profit maximization and purely rational decision making, while not including emotions to their full extent, is as embedded in Heckhausen's approach, as it was in the earlier discussed expectancy x value theories.

The mainly profit oriented approach has also been stated by Tversky (1975, p.5): *"utility theory does not permit attitudes toward risk per se, only attitudes toward money"*. Furthermore, Björkman, (1984), Mitchell (1982), Fischhoff, Goitein and Shapira (1982) and Lopes, (1984) all pointed out that prospect theory, as the other expectancy value theories, does not represent daily decision making under uncertainty, or risk taking, but only deals with some well defined situations (such as studies on experimental gaming), where both the time and emotional component of decisions and risk taking are omitted completely. As, however, emotions are strongly involved in decision making and risk taking, a complete theory of decision making should include them (Lopes, 1987).

4.3.6 Two-Factor Theory for Risky Choice

Lopes (1987) proposed a two-factor theory for risky choice, in which she combined personality and situational factors. Lopes (1987) introduced her security/potential personality factor by describing risk-averse people as having a desire for *security*, while risk seeking people have a desire for *potential*. The situational factor is called *Aspiration level*, and refers to the opportunities at hand for the decision maker. The aspiration level can be set by external influences, choice characteristics, internal cognitions, expectations, and emotions. The factors also interact to produce a final decision in risky choice situations. Although a dispositional factor and a situational factor are introduced, the theory is still largely focussed on rational, conscious decisions, thereby neglecting emotional, physiological and sub-conscious components of risk taking. However, Lopes (1987) stressed the importance of these aspects and tried to include emotions as a form of rational decision making, such as stating that planning is applied hoping.

4.3.7 Affective Balance Theory

A theory that attempts to incorporate emotional and physiological components of decision making is affective balance theory. Grossberg and Gutowski (1987, p.301) stated that it deals with:

"phenomena concerning human decision making under risk to human evoked potentials, neurophysiological and pharmacological substrates of behavior animal discrimination learning, human memory and attentional processing, and certain mental disorders"

The theory postulates that affective comparative strategies in decision making are not explicitly determined by the stimuli. The cognitive context may alter the interpretation and consequent behavior. The process regulating affective or emotional experience is an *opponent process* called *gated dipole* (A control mechanism selecting input/output in both directions). The selection works through chemical transmitters, competitive, opponent interactions between an on-channel and an off-channel, and phasic inputs that perturb these channels over time. A sustained, or tonic arousal level perturbs both channels and thereby sets the sensitivity of dipole outputs in relation to phasic input fluctuations (Grossberg & Gutowski, 1987).

The theory suggests that high probability events are more easily accessed in both the short-term and long-term memory. Specifically, it postulates that:

"activation of an event's short-term memory representation elicits signals from this representation to the gated dipole opponent process where emotional reactions are generated".

Before these signals can reach the gated dipoles, they are multiplied, or gated, by long-term memory traces encoding the conditioned reinforcer values of the event. These gated signals activate either the on- or the off-channel of their target dipoles, and thereby triggers emotional reactions and motivational signals (Grossberg & Gutowski, 1987). Relative to SEU theories, affective balance theory offers similar assumptions about psycho-physical value functions, decisional weights and probability-event-value information integration rules. A solely rational, conscious-attention driven comparison between affective changes, responses and conditions is proposed. The input information is broken down into pairs of alternatives, which are then evaluated rationally, combined with their affective value.

Shortcomings of Grossberg's approach lie in the assumption that decision makers are solely rational, - even regarding emotions. Furthermore, the theory cannot accommodate the taking of risks for the physiological and emotional pleasure of experiencing arousal. However, the attempt to explain decision making under risk by an emotion regulating process represents a major step forward in the application and development of utility theories (Grossberg & Gutowski, 1987).

4.3.8 Decision Making as a Product of Evolutionary Theory

In an attempt to reduce the uniqueness of decision theories to a mere derivate of evolutionary theory, Cooper (1987) argued that decision logic, utility theories and subjective probability theories can be just as easily described as developments of evolution. The author argued that thinking in terms of utility maximization was a product of the survival of those organisms which developed the brain structure to apply exactly these heuristics. He argued that the objective probability of surviving the use of a specific decision structure increased the likelihood of offspring and therefore made organisms who use this structure better equipped for survival. Cooper (1987) explained decision theoretical models by adopting an evolutionary approach. For him, subjective probability theory is a product of objective probability theory combined with evolutionary biology. Although Cooper (1987) admitted that his explanation is far too simplistic, the notion that decision heuristic structures, or at least a disposition to develop some structures and not others is quite logical and conceivable. It is also in line with modern socio-biological theories, as proposed by Wilson (1975), Dawkins (1976; 1986), or Crawford and Anderson (1989). The degree of the disposition, or the development of heuristics could also serve as a starting point for incorporating individual differences into decision making theories. Different individuals may choose different heuristics because their neural network is predisposed to acquire exactly these heuristics easier, while the environment determines the applications (Crawford & Anderson, 1989).

4.3.9 Multi Attribute Utility Theory (MAUT)

Another approach is the use of Multi Attribute Utility Theory (MAUT), as proposed by Keeney and Raiffa (1976) and adapted to risk by Fischhoff, Watson and Hope (1984). The basic assumptions of MAUT are given by Fischhoff et al. (1987, p.56):

"A structured form of simplification is offered by techniques such as multi-attribute utility theory, which decompose complex questions into more manageable components, each of which considers a subsidiary evaluation issue. The structuring of these questions allows their recomposition into overall evaluations, which are interpreted as representing the summary judgement that respondents would have produced if they had unlimited mental computational capacity."

The advantage of MAUT lies in the ability to split complex processes and decisions into easier parts, which may resemble the decision process we undergo when faced with a complex problem (Keeney & Raiffa, 1976). Fischhoff (1977) further stated that MAUT makes tradeoffs more explicit and puts attributes of unequal material values on a comparable basis. The disadvantages lie in its limitation to areas where only simple relations exist, its vulnerability to errors in interpretation, because individual components may point to a different conclusion than a holistic analysis, its neglecting of interactive components, which denies the possibility that factor A may influence factor B differently at different levels, and finally in its tendency to direct subjects answers to follow the split of the researcher (Fischhoff et al. 1987). Overall, it is likely that every complex decision is made out of a subset of individual utility assessments and their combined value (Keeney & Raiffa, 1977). The procedure, however, cannot be readily used in many areas (Fischhoff et al., 1987). It is also doubtful whether complex human interactions can be seperated into meaningful components without losing information about the process as a whole.

4.3.10 Decision Field Theory and "Random Models"

Busemeyer and Townsend (1993) put forward that decision theory accounts for the motivational and cognitive mechanisms that guide the *deliberation process* involved in decisions under uncertainty. They argued that deliberation is a time-consuming and effortfull process manifested by indecisiveness, vacillation, inconsistency, and distress. Busemeyer and Townsend (1993) further stated that this component of the rational decision process has not been and can not be accounted for by the above reviewed utility theories as the problems faced when making complex decisions are dynamic and the theory accounting for it should be stochastic in nature. They argued that probabilistic assessments and analyses have to be performed to predict behavior more accurately as either:

1) *Deterministic Subjective Expected Utility (SEU),*
2) *Random SEU Theory,*
3) *Sequential SEU Theory, or*
4) *Random Walk SEU Theory.*

Decision field theory is assumed to incorporate the descriptions and predictions made by these and similar theories as they can be accounted for more accurately through probabilistic calculations. Busemeyer and Townsend (1993) do, however, admit that their main focus is on deliberate, rational decision making and not on automatic "instinctive" irrational or other forms of decisions.

English and Allison (1993) went one step beyond the generally agreed upon axioms of SEUs, namely that people try to maximize benefits and minimize costs. Based on economic assumptions they argued and demonstrated mathematically that random models can explain at least equally well and sometimes even better, empirical phenomena typically found in SEU experiments. They argued, that instead of treating all deviations from the *rationality axiom* as imperfect control and assessment on part of the individual, other models and assumptions may serve better to explain human behavior under conditions of uncertainty. The minimum requirement for a decision making model should be do make better predictions than a random model.

4.4 Utility Theories and Risk Taking

Although formal descriptions of utility, or cost/benefit theories are limited in their applicability to day-to-day life decisions, they offer a well developed tool to understand conscious and rational decision making under conditions of uncertainty. If subjects have enough time, understand the problem and approach it by rational (maybe only subjectively rational) thoughts, decision theory can predict behavior relatively accurately. However, to understand spontaneous, not-conscious, emotional, or physiological components of risk taking behavior, these theories have to be extended substantially. Fischhoff et al. (1987) suggested, that exposure to risk itself can bring benefits, either in form of enjoyable arousal and intensified emotional experiences. They suggested to put emotional and material benefits into a "net benefit" assessment of all benefits and non-risk costs.

Larrick (1993), and Josephs, Larrick, Steele, and Nisbett (1993) showed that one psychological factor usually not incorporated into expectancy x value assessment is that of protecting one's self-image from the consequences of a wrong decision. Furthermore, the expectation of negative consequences seems to be dependent upon the subjectively "feared" damaging value of those consequences for the self-image. Again, although these authors demonstrated the importance to incorporate intra-personal, emotional consequences into risk taking behavior, they fail to account for the attractions and positive benefits (such as becoming a "hero", or the saviour of the company) that risky decisions undoubtedly also possess (Clifford, 1991).

Portfolio theory (Coombs, 1972) and Fishburn's (1970) risk preference theory are in their "optimizing" assumption similar to optimal level of arousal theories. A combination of personality and utility theories is presented by Rofé (1984). He reported that the strength of an affiliation tendency depends on a cost/benefit analysis, which considers all possible outcomes of affiliation. Situational determinants, such as the perceived dangerousness, controllability, affective nature of the situation or its outcome (e.g., embarrassment) were mentioned as influential in the decision whether to affiliate.

Furthermore, Rofé described personality characteristics, such as being a "repressor", thereby avoiding, denying, filtering perceptions to avoid thoughts about potential dangers, as opposed to "sensitizers", who showed the opposite behavior (Rofé, 1984). They found empirical support for this distinction in pregnant women, who preferred company to reduce the fear when they were sensitizers, as assessed by questionnaire, compared to repressors who showed less of an affiliation tendency, since they repressed the presence of fear producing factors.

After discussing utility theories and their relation to risk taking behavior it seems to be a logical conclusion that risk taking behavior is not motivated and controlled completely through rational decision processes. The major proponents of such decision making approaches towards risk taking do admit, that much of our day-to-day behavior is based upon preferences, ideas, motives and decisions that are in violation of the utility maximization axioms. Whether they may be explained by including emotional, physiological or other non-material motivational and emotional processes, will be discussed in the next chapter.

Chapter 5: How Is Risk Taking Motivated and Emotionally Experienced?

Most research dealing with risk taking (e.g. Yates, 1992) relates to rational, mainly materialistic and conscious motivations for risk taking. Goal-oriented expectancy x value utility concepts do, however, only account for some of the many facets of risk taking. Another motivating facet is that of sensation seeking, which has been discussed in chapter three. Yet another aspect of risk taking originates in the motivation to be successful in mastering dangerous and difficult tasks or in achieving something extraordinary in life. These aspects are discussed partly in theories dealing with achievement motivation. A major motivating force for risk taking has been identified as positive emotional experiences. Thus, the relationship of emotions, cognitions and risk taking will also be discussed in some detail in this chapter. Finally, many risk taking situations, such as driving in a car, walking on uneven surfaces, or talking about important issues incorporate appraisals, schemata and behaviors which are executed below the threshold of consciousness. I shall begin with a discussion of motivation, especially focussing on achievement motivation and risk taking.

5.1 Risk Taking and Theories of Motivation

An intrinsic reward system for reaching desired goals was postulated and has found empirical support (see chapter three). Thus, striving to obtain a goal should also be intrinsically and genetically motivated, because the achievement of goals, such as obtaining food and shelter would be advantageous from an evolutionary standpoint. It would motivate people to take certain risks. The fear of failing to achieve one's goals would offer the motivation to be more cautious and control one's environment. Similarly to the argument presented before, namely that risk taking behavior is normally distributed and subject to individual differences, I shall argue the same for achievement motivation. Risk seekers and risk avoiders, who are located at different ends of the achievement motivation dimension (Atkinson, 1957), are motivated by differently difficult situations, and are intrinsically and extrinsically motivated to pursue their respective risk taking activities (see also Deci, 1975; or Csikzentmihaly, 1985; 1988).

Intrinsic motivation is defined here as: *performing an activity for pleasure derived from the execution of the activity itself,* while ***extrinsic motivation*** *is triggered by the anticipation of external rewards that may follow the successful task completion* (Matsumoto & Sanders, 1988).

In the chapters above, I provided support for the assumption that risk seekers and paratelic individuals pursue endeavors that are often neither goal-oriented, nor material-oriented, that are not consciously decided upon, or predictable and stable. Risk avoiders and telic individuals show the opposite preferences than paratelic individuals. Furthermore, some situations favour one type of risk seeking behavior more than others. Thus, people will be motivated to engage in the subjectively most rewarding type of activity. As the literature on motivation and achievement is incredibly voluminous, I shall only focus on some of the research especially related to risk taking.

I shall follow the historic development for achievement motivation, such as Lewin's (1938; 1951) basic work, Atkinson's (1957) study on the relationship between achievement motivation and risk taking, McClelland (1951; 1961; 1985) and his personality/situation interaction model, and Weiner's (1986) attributional model. Atkinson and Birch's (1978), and Blankenship's (1986; 1987) Dynamics-of-Action-Model will be followed by Dweck and Leggett's (1988) personality approach. Then the connection between risk taking, motivation and emotion will be discussed.

5.1.1 Risk Taking Behavior and Achievement Motivation

Lewin (1938; 1951) presented the *Level of Aspiration Theory* as a choice situation determined by values of success and failure and the probability of future success. All components are weighted individually and the resulting aspiration level is the motivation for behavior. In the level of aspiration theory success and failure do not have to be inversely related, they can also occur simultaneously, thereby accounting for both ambiguity and aspirations with outcomes of mixed expected values. This utility theory of motivation could thus be viewed as the basis for explaining positive and negative motivations of risk seeking and risk avoiding.

Atkinson's (1957) *Theory of Achievement Motivation* stated that achievement oriented, risk seeking subjects preferred problems of medium difficulty, since these offered them the best chance to prove their abilities. Risk avoiding, fear-of-failure oriented subjects preferred either very easy tasks, that ensured success, or very difficult ones that provided them with an excuse for failing. Atkinson (1957) found that risk seeking subjects preferred a chance level of 50%, which was least liked by risk avoiding subjects. The motivation for risk taking behavior is a multiplicative function of the strength of the motive, the expectancy (subjective probability) of success, and the value of the incentive. Studies by McClelland (1958), Atkinson et al. (1960) and Kuhl (1978) supported the theory by observing correlations between preferred difficulty in a ring-toss game with achievement scores.

Clifford (1991) examined the hypothesis of a preferred 50% probability of success and reported that many studies and his own empirical research show numerous violations of this assumption, mainly attributed to the subject's wish to accomplish something extraordinary and thus choosing higher risks. Sansone, Weir, Harpster, and Morgan (1993) showed that subjects transform regular, so called "boring tasks" such as marking exams or learning passages by heart into more challenging ones by inventing imaginary competitions (pages per hour) or setting task-independent goals (finding the most creative misspellings). Apparently the need for interesting and varied stimuli seems to motivate novel and sometimes even cost-intensive behavior patterns. Thus, the factors involved in the motivation to take risks are apparently not only cost/benefit assumptions (minimax strategy) but rather depend on non-material psychological factors such as self-esteem, feelings, and anticipation thereof.

McClelland's (1951; 1985) personality concept is also built around the need for achievement concept, although with a stronger focus on cognitive, rather than "drive" components. McClelland (1985) also stressed the importance of individual differences, which can be found in the personality traits related to hope-for-success versus fear-of-failure in his *Need for Achievement Theory (n-Ach)*. McClelland (1985) further stated that physiological and drive aspects, as well as emotional experiences can influence every achievement motivated behavior, so that the complete turn towards a purely cognitive model was premature. McClelland (1985) argued that instead of limiting achievement motivation behavior to external rewards, innate reward structures for achievement motivation in the mid-brain (similar to those suggested to trigger arousal) should also be considered.

Weiner (1985; 1991) focussed on attributional aspects of achievement motivation. He stated that from ancient times on, everybody was constantly trying to understand causal principles for certain phenomena. Such causal principles are necessary for the previously mentioned approaches of predicting, understanding and mastering a steadily changing environment, in order to increase the likelihood of one's survival. Similarly, Weiner (1986) argued for the importance of variables, such as personal controllability of cause, locus of the cause (related to locus of control), and stability of causes. These dimensions are accompanied and partly regulated by emotional experiences of anger, guilt, pity, pride, gratitude, hopelessness, and shame. Furthermore, emotions are also elicited by experiencing a positive outcome, regardless of the attribution, but especially, if success can be attributed to oneself. Weiner (1986) further stated that anger, pity, guilt, and gratitude are important for maintaining social order and promoting gene survival, thus opening motivation research to a neglected area.

5.1.2 The Dynamics of Action Model of Motivation

Atkinson and Birch (1978), and later Kuhl and Blankenship (1979), and Blankenship (1986; 1987) developed Atkinson's (1957) theory into a *Dynamics of Action Model*, to account for how achievement motivation changes over time towards more difficult goals, even if the probability of outcome is held constant. Subjects high in need for achievement and low in test anxiety initially choose tasks at a medium risk-level (probability of 0.5) but then shift towards more difficult goals, with lower probabilities of success. This tendency towards a more risky behavior cannot be readily accounted for by Atkinson's (1957) stable model. Within the dynamics of action model, however, the risk-activity choices depend on the interaction of instigation, inhibition and consummation.

Instigation results from past reward experiences and is equated with hope for success, while inhibition is associated with past punishment experiences and is equated with fear of failure. Consummation (satiation) is associated with success/failure outcome. It explains the finding that the motivation reduces in strength after the activity has been performed (similar to drive reduction) (Kuhl & Blankenship, 1979; Blankenship, 1987).

Motivation changes in dominance, depending on the stimulus, outcome, and association interactions, and the strongest motive gains control of the behavior (Blankenship, 1986). The finding that subjects preferred the more risky action after choosing an intermediate difficulty, can now be explained by the consummatory tendency decreasing the motivation to stay at an intermediate level, so that the instigation tendency motivates subjects to engage in more risky behavior (Blankenship, 1982).

In other words, as mastery increases the degree of uncertainty, and thus the level of arousal decrease. To prevent the level of arousal from dropping below the preferred one the subject chooses a more difficult task, or more difficult items. In a study on the effect of anxiety in a computer task and its effect on achievement motivation, Blankenship (1987) found that subjects low in need for achievement required about three times as long to cope with the anxiety producing breakdown of the computer, before continuing with the original task. This procrastinating behavior of subjects low in need for achievement let Blankenship (1987) deduce that people with low resultant achievement motivation and high anxiety avoid engaging in subjectively risky choices, until they are provided with an excuse for possible failure (such as no time to act properly), or the fear of failure in the required task urges them to engage in it (loss of money, prestige, or other rewards).

Schneider and Posse (1982) found support for the dynamics of action theory in a psychomotor task, and in a computer simulation. In addition, they offered an alternative explanation in that subjects just go from the easy to the difficult task to test at what difficulty they can still be successful. Changes towards riskier choices are thus only an attempt to learn more about the task and their abilities. Meyer, Folkes and Weiner (1976) argued for a general tendency to understand the environment. Thus, subjects choose an intermediate level of risk. They found that most subjects chose intermediate difficulties, regardless of their need for achievement score.

The notion of testing one's abilities and mastering one's life as a motivator for engaging in risk taking behavior was also reported by Piët (1987) in a study on stuntmen. These subjects were all extremely high in need for achievement orientation in general, and in the desire to be noticed in particular. The subjects were also high in challenge seeking, low in anxiety and low in their boredom threshold. The combination of risk seeking and desire for mastery and control makes them engage in risky jobs, while simultaneously maximizing their personal control over the outcome. Subjects reported being emotionally, physically (adrenalin), and cognitively rewarded for taking the risk, as well as for mastering it. The motivation to achieve in risk taking activities is therefore very strongly reinforced.

5.1.3 Individual Differences in Risk Taking Motivation

Dweck and Leggett (1988) argued that cognition-affect-behavior patterns can be split into a "helpless" response, and a "mastery" response. The helpless response is characterized by an avoidance of challenge and a deterioration in the face of obstacles, while the mastery response pattern involves the seeking of challenges and the maintenance of effective striving under failure. Thus, mastery-oriented people are similar to people high in the need for achievement, sensation seeking-oriented and high in desire for personal control. Furthermore, Dweck and Leggett (1988) argued that people with the "helpless" response pattern are also striving for activity oriented goals. "Performance goals" are generally pursued by people who are concerned with gaining favorable judgements of their competence. "Learning goals", on the other hand, are more important for subjects interested in increasing their own competence. This tendency is ascribed to people showing the mastery-oriented response pattern.

In addition, Dweck and Leggett (1988) stated that those people in general, and children in particular, who follow a "helpless" response pattern and are activity oriented are characterized by a stable, unchangeable personality picture, which lets them attribute failures to personal inability. Every problem is viewed as a threat to the self-esteem, while "master" people perceive problems as challenges to learn and prove their abilities (Diener & Dweck, 1978; 1980). Nicholls (1984) split mastery into either compared to oneself or to others. Different attributions for personal ability result from choosing either approach. A gain in mastery is only perceived as rewarding when it is self-referenced. If an ability is compared to that of others, one has to reach equal achievements with less effort to experience the emotion of pride.

Elliott and Dweck (1988) showed that even highly confident, performance oriented individuals avoid situations with high risk of failure. Dweck and Leggett (1988) further argued that learning oriented, "helpless" people have a tendency to show strong conformity to social standards and are generally risk avoiders. They attribute causes to external rather than internal control, and perceive their personality, intelligence and chances in life as less under their personal influence, than "mastery" individuals, who tend to compensate for less personal influence over a situation by increased effort (Dweck & Leggett, 1988).

Dweck and Leggett (1988) also stated that dispositions towards a certain type of response are important, but that situational influences can override these dispositions. Thus, they suggested that personality theories should not just predict the same behavior in every situation, but rather predict what particular behavior will be displayed in which situations. The interaction of personality, situation, cognition, emotion, and behavior produces a motivation for people to engage in challenge seeking and risk taking behavior. Tasks, such as going to school and having to learn certain material is usually not liked very much (Sansone et al. 1992). Parker and Lepper (1992) showed, however, that presenting learning material in a fantasy context that raised children's interest and imbedded the otherwise boring task into a challenging, fantasy and creativity stimulating context (e.g. in a computer game) produced significant increases in performance. Risk taking and challenge seeking may thus be utilized in underexplored domains, such as academic learning motivation.

5.1.4 Academic Risk Taking Motivation

Clifford (1991) reviewed the literature on academic risk taking and reported that people in general tend to seek out challenging and risky situations. Most studies on achievement motivation do not reveal a preferred 50% probability rate of success (as suggested by Atkinson's theory) but rather varied ones, ranging from 20% to 85%, depending on the subjective value of succeeding and personality factors alike (Harter, 1978). Students seem to acquire over time a more risk avoidant tendency as they seem to have learned that every creative, daring deviation from normative behavior and answers is punished or at least not rewarded. Clifford's extensive research showed, however, that adults in general and students in particular can be stimulated and motivated towards better learning and understanding of class material by transforming regular tasks into risk taking ones. Reasons for the success of risk taking, challenging classroom activities are seen in:

- increased self-efficacy (Bandura, 1977)
- optimizing cognitive development (Clifford, 1991)
- more complex strategy orientation and thus learning from failures
- increased need satisfaction as risk taking is a universal tendency
- increased and accurately measured competency through feedback
- self motivation and flow-experiences (Csikszentmihalyi, 1988).

Clifford (1991) concluded that psychological theory and empirical evidence are in stark contrast to public opinion and especially to beliefs dominating many educational systems (p. 290):

"Whereas motivation and cognitive developmental theories tells us that moderate to high levels of challenge are needed to achieve learning and skill development, popular opinion tells us to minimize error making during the learning process, or academic achievement will be retarded.
Theory tells us that errors, cognitive desequilibrium, and failure to reach one's goals are major sources of continued motivation; popular opinion tells us that failure wreaks havoc to student motivation and undermines self-esteem. Motivation theory tells us option, choice, and self regulation of behavior are essential for optimum motivation and performance; public opinion tells us that student motivation and achievement are best attained when incentives and reinforcers are used to elicit student compliance with teacher-determined assignments and activities.

Her plea is to apply empirical and theoretical finding to education by stressing positive, challenging aspects of risk taking in the classroom instead of imagining "schools without failures". Clifford (1991, p.293) suggested that instead of preventing or limiting students` potentials the following means should be used to support creativity and self-determination:

1. *transform assignments and review activities into risk taking activities,*
2. *manage the frequent use of risk taking activities and ensure the provision of informational feedback,*
3. *establish multilevel competence criteria for risk taking activities,*
4. *identify practical means of measuring and monitoring learning from risk taking (e.g., the correction of errors and changes in cognition),*
5. *modify traditional reinforcement programs so that they encourage rather than discourage risk taking, and*
6. *create classroom environments conductive to academic risk taking.*

That way students will learn when and where to take risks, assess them correctly and enjoy learning and exploring rather than dread it.

5.2 Risk Taking and Emotions

In this chapter I shall argue that emotions are genetically programmed motivators for risk taking behavior. The function of emotions can be better understood when viewed from an evolutionary standpoint, as they have had an important survival value (Levenson et al., 1990). I shall further argue that the debate whether emotions precede cognitions is essentially unproductive and outdated, since emotions and cognitions are part of the same process, with conscious and not-conscious appraisals. I shall discuss the relevance of Frijda' s (1988) laws of emotion to risk taking behavior and the concept of Izard (1993).

5.2.1 The Motivating Role of Emotions in Risk Taking

Toda (1980, pp. 136-137) quoted from the memoirs of the former president of the USA, Truman, the following decision process, about whether to take the considerable risk of engaging in a war:
"In deliberating whether or not to resist the invasion of South Korea, President Truman reports considering how he would feel about himself if he, as President, failed to resist the aggressive action of another nation."

Toda (1980) further brought examples of our everyday life risk taking decisions, such as whether to go on vacation to this or that place, or whether to have a date with someone, or whether one should take a risky or safe corporate decision as a manager. Lazarus (1991) suggested emotions to be major motivating goals in our day to day lives and to give importance to situational considerations. In all these examples the influence of emotions on risk taking is very obvious and can experienced by all of us. Few people deny anticipating joy with a vacation, or feeling attraction towards a dating partner, or thinking about pride versus shame in important business decisions. Toda (1980) even argued that emotional motivations are often far more important in critical moments than material ones. Oatley and Jenkins (1992) reviewed the functions of emotions and came to the following conclusions:

- Emotions are mental states and mental processes that can be measured behaviorally, physiologically and by self-report
- Emotions do not have a consistent definition in the literature
- Emotions serve as communicators to others and oneself
- Emotions serve to insert and maintain information about events
- Emotions serve to maintain information about consequences
- Emotions can rearrange the priority of goals and actions (Isen, 1990)
- Emotions are often conscious (Lazarus, 1984) but do not have to be present or active in the consciousness of an individual (Lang, 1988; Zajonc, 1980; 1984)
- Emotions do have a positive effect on creative problem solving (Isen et al., 1987)
- Emotions do have a significant influence on physical and psychological well-being (Oatley & Jenkins, 1991)
- Emotions influence organizational behavior, including helpfulness, managerial risk taking and negotiating (Isen & Baron, 1991).

Frijda (1988), and Frijda et al. (1989) see the experience of pleasure and pain at the core of emotions, while Lazarus (1991) defined pleasure and pain as innate reflexes rather than emotions. Emotions follow certain neurological pathways (Murphy & Zajonc, 1993), action patterns, and laws, regardless of their particular type (Frijda, 1987). Two theories will be presented which combine several aspects of the risk taking-emotion link.

Prime Theory

Buck's (1985) *Prime Theory* bridges the gap between biological and cognitive theories of motivation via the use of emotions. Buck (1985) argued that the finding that animals anticipate homeostatic deficits in their needs (for example eating so as not to become hungry) speaks strongly for an emotional reward system that motivates the organism to a certain behavior, before a drive reduction tendency becomes necessary. Based on Darwin (1872/1934) and Izard (1977), Buck (1985) argued that emotions are an evolutionary requirement for survival, necessary for learning, and a motivator for behavior.

Buck (1985, p.390) defined ***"prime"*** in prime theory as:
"biologically based primary motivational/emotional systems (primes) have evolved within each species with the basic role of bodily adaptation and maintenance of homeostasis".

Buck (1985) based the importance of primes on evolutionary developments. He argued that each species has developed unique ways of responding to challenges through adaptive behaviors of approach and withdrawal. Some of the systems are present in all organisms, such as need for food and water, while others are highly species specific, such as migration for birds, and long attachment to the young in primates.

Primes work on all levels of brain organization. In the lower levels of the brain, (brain-stem, midbrain, and basal ganglia), which are important for the regulation of autonomous regulation, such as heartbeat, blood pressure, breathing, little if any emotional regulation can be detected, while the limbic system and hypothalamus, as well as the amygdala in the intermediate levels of the brain, clearly show emotional involvement, both in form of automatic, reflex-like emotions, as well as in fixed action patterns, and in the basic emotions, or "prime affects" of happiness, sadness, fear, anger, surprise, and disgust. The highest level of the brain, the neocortex, is involved in complex emotional-cognitive interaction patterns and deliberate, strategical thinking.

Buck (1980; 1985) argued, that although some stimuli elicit emotions inherently, and automatically, for example the emotion of anger, the way this anger is expressed and processed later on is individual specific and learned. The autonomic response of the adaptive/homeostatic system, however, begins regardless of our conscious interpretation. It may cause physical stress with the shutdown of the immune system as a result of the anger-arousal perception. Brain-internal reward systems (identified by Olds & Fobes, 1981), combine automatic bodily responses with perceived emotions, such as happiness and satisfaction.

Buck (1985, p.394) defined motivation as:
"Motivation is a potential, inherent in the structure of the primes. Motivation is a general term for the forces controlling-activating and directing-behavior. It is defined here as a potential that is inherent in the structure of systems of behavior control".

This motivation potential of the primes becomes automatically active when a challenging stimulus elicits the evolutionary advantageous emotional response within the organism, just as energy manifests itself in matter. Both energy and matter, as well as emotion and motivation are partly the same, but serve different functions, and have different appearances. Consequently, Buck (1985) argued that emotions are constantly present, either pleasantly or unpleasantly, but do not have to be consciously experienced, and are often ignored until they change. For example, driving at a certain speed for a longer time may not be consciously perceived as emotionally pleasant or unpleasant. Then, as soon as the speed is perceived as unpleasantly boring, or anxiety provokingly fast, the emotional reaction, and its cognitive evaluation, immediately become conscious.

If applied to general risk taking behavior, prime theory would suggest that the initial stimulus is already experienced on a basic physiological-emotional level, and leads to autonomous, automatic bodily responses. Higher levels of emotional experience are still very subjective and underlie learning aspects, as well as the cognitive-emotional interactive process.

It has been argued that risk taking is most likely an emotionally rewarded behavior. Thus, the combination of an initial experience of pleasantness in response to a challenge and the resulting arousal, will lead risk seekers to interpret the situation as a stimulating, positively challenging one and prompt them to look for ways to master the situation. Risk avoiders, on the other hand, will probably perceive the initial stimulus as an unpleasant threat. They will then respond with a cognitive evaluation that the situation should be abandoned, or not to be entered at all. The subjective emotion-conditioning history will thus determine risk avoiding behavior.

Different situations will therefore elicit different associated emotions, and different individuals will respond differently to them. The spectrum of emotional experiences in risky situations will vary greatly within a population, but be constantly present. Prime theory combines elements of evolution, physiological arousal, motivating emotions and evaluating cognitions to account for many postulates about risk taking behavior.

Frijda`s Laws of Emotion

Complex emotions require a more cognitive and evaluative processing of emotions. Frijda (1986; 1988) reviewed some of the research on emotions and summarized the findings, as well as the postulates still in need of testing in form of laws. He argued that the laws of emotion are grounded in mechanisms that are not of a voluntary nature and only partly under voluntary control. Frijda (1986) postulated that emotions involve a change in *"state of action readiness"*, which consists of the following three components:

a) readiness to approach, withdraw, focus attention,
b) sheer excitement, and
c) being stopped, or losing interest.

Frijda further stated that we obey the following laws of emotion virtually automatically, since we cannot enact them at will.

1. ***The law of situational meaning*** refers to the finding that different situations elicit different emotions. Unpredictability and uncontrollability seem to be key contributors to the shaping of this law (Mineka & Henderson, 1985), which have also been identified as important factors for risk taking, risk assessment and risk acceptance.

2. ***The law of concern*** expresses the finding that emotions arise in response to events important to the individual's goals, motives, or concerns. Frijda argued that emotions arise from the interaction of situational meanings and concerns. The law is supported by risk homeostasis theory (Wilde, 1982; 1986) which states that personal values, motives and concerns determine the target level of risk, and therefore the risk taking behavior.

3. ***The law of apparent reality*** refers to emotions being elicited by events appraised as subjectively real, in which the intensity corresponds to the degree in which this is true. The central element of the law is, that emotions do not respond to "objective" reality, but to our subjective experience of it. For example, we get angry when someone steps on our toes, even if we know the other person is not to blame. We do not perceive objective risks, but our subjective estimate of it. If we do not perceive the risk (e.g., radiation, invisible gases) we do not act, and if we believe a risk to be present, we respond, regardless of whether the risk is really there (e.g. phobias).

4. ***The laws of change, habituation and comparative feeling*** refer to the elicitation of emotions due to changes in external conditions, and to the comparison of some frame and reference, such as in *"Schadenfreude"* versus *envy*. The laws suggest that changes in perceived riskiness will elicit corresponding emotional experiences.

5. ***The law of hedonic asymmetry*** refers to the effect that continuous pleasure wears off while pain persists, leading to processes of self-deceit and self-defense, and to a continuing search for new sources of pleasure, especially for sensation and challenge seekers.

6. ***The law of conservation of emotional momentum*** refers to the finding that time does not heal all wounds, but constant cognitive exposure may do. Traumatic experiences, such as the loss of a child, however, may always preserve their emotional momentum. This law may explain some of the rewarding character of emotions in repeated risk taking tasks, such as racing around a track, or betting in gambles.

7. ***The law of closure*** refers to the observation that emotions are not relative and deliberative, but close off the individual for other considerations. The law of closure often takes absolute control precedence over the situation and determines the action more than any deliberate reasoning. This may explain why people do close themselves off when presented with pictures of dreadful accidents.

8. ***The law of care for consequences*** is another law referring to the influence of personal control over the future. It postulates a compensatory process to balance emotional costs and benefits.

9. ***The laws of lightest load and greatest gain*** refer to the desire to maximize emotional benefit and minimize emotional load.

Complex emotions and appraisal patterns have been the focus of many studies examining the processes and relationships between cognition, attribution, and emotion. For example, Roseman, Spindel and Jose (1990) examined the effect that particular appraisals had on eliciting emotions and found varying patterns of emotions depending upon the appraisals of:

- Situational state (event consistency with motives)
- Probability (of outcome)
- Agency (locus of cause, self/other)
- Motivational state (reward vs. punishment orientation)
- Power (coping potential, controllability).

Similar dimensions and laws of appraisals with emotional consequences but with partly different names were found by Smith and Ellsworth (1985; 1987), and have been further discussed in their relationship to cognitive processing (e.g. attributional processing) by Weiner (1985) and to primary versus secondary appraisal in Lazarus (1991). Such *laws* are meant as possible explanatory factors for the influence emotions have on our behavior, especially risk taking behavior. However, significant individual differences in responsivity, coping, experience, or evaluations of emotions have been reported in the literature.

5.2.2 Individual Differences in Affective Reactivity

It has been shown in the chapters above that risk taking is related to physiological arousal, and emotional experiences. Furthermore, people adjust their risk taking-arousal experience to a subjectively optimal level, accompanied by an emotional well-being. Emotional reactions towards life-events differ in people relative to their physical arousability (Larsen & Diener, 1987), and people start at different baselines of arousal (Eysenck, 1967; Zuckerman, 1979) with individually different vicarious emotional responses (Eisenberg et al., 1991).

Apart from the influence of arousal on emotions, Larsen and Diener (1987) suggested that the intensity of an individual's affective responses operates as an automatic compensatory mechanism regulating the individual's level of physical arousal. This assumption forms the basis for arousal regulation theory (Emmons & King, 1989). Furthermore, Larsen and Diener (1987) and Emmons and Diener (1986) suggested that an individual's tendency to experience extreme and variable emotions may mediate the complexity or differentiation of the individual's cognitive system. Emmons and King (1988; 1989) further deduced that people set their goals, design their social relationships, and structure their concept of self-representation in such a way that it serves their affective reactivity best.

In other words, according to Larsen and Diener (1987) individuals with high emotional reactivity seek out and design their lives in such a way that it offers them a wide range of emotional experiences and arousing situations, which in turn make them feel best. Being exposed to emotion arousing situations creates expectations, goals, and cognitive patterns influencing future behavior (Emmons & Diener, 1986). The earlier in childhood these patterns are formed, the stronger they will influence the personality patterns (Emmons & King, 1989).

Empirical support for these assumptions was reported by Larsen (1987), showing that affective intensity was related to a high frequency of mood changes over time and that aroused subjects showed a preference for strong emotions and emotion-provoking situations. Emmons and King (1989) demonstrated that subjects who were emotionally more reactive and displayed a wider range of emotional reactions were also more differentiated in their goal-striving systems. The authors explained their results as being in accordance with the postulate of people actively seeking out, creating, and influencing their environment in accordance with their psychological motivations and desires (see also Emmons & Diener, 1986).

The view that emotions regulate and are being regulated by arousal, and that cognitive processes and personality types are being controlled by the processes of arousal and emotional reactivity fits well into models of sensation seeking, risk homeostasis and models postulating intrinsic reward systems for the striving for emotional well-being. Thus, emotions and an optimal level of arousal serve as motivators for risk taking.

Buck (1980) stressed the importance of feedback for the development and use of emotions. Visceral feedback has been shown to influence emotions, and proprioceptive feedback seems to be essential for the experience and maintenance of the expression and experience of emotions (Izard, 1977; 1984). Some indices have been found relating neural feedback about emotions in the hypothalamus to cognitive evaluations and memories.

Stepper and Strack (1993) showed that success at an achievement task led to greater feelings of pride if the outcome was received in an upright body position rather than in a slumped posture. Also, contraction of the forehead muscle led to greater perceptions of effort. These findings show, that physiological states and conditions can have a direct influence on emotional experiences, and lead to different cognitive appraisals and evaluations.

Buck (1985) pointed out that large individual differences in the experience and expression of emotions are likely. Larsen, Diener and Emmons (1987) reported substantial individual differences in emotional arousability and emotional experiences. They compared their concept to the strength of the nervous system approach (Pavlov, 1927) and the reducer/augmenter personality types (Vando, 1974). The learning experience from different cultures and different groups of the same society led to the development of different emotional patterns, the understanding of which could support emotional education within any society (Buck, 1985).

Ekman et al. (1985) discussed individual differences in emotional expressions. The only emotional expression showing no individual differences across people is the startle reaction, a behavior experimentally induced, for example, by the surprising firing of a pistol (Ekman et al., 1985). The same authors doubted the status of the startle reaction as an extreme surprise emotion, while other authors (Tomkins, 1962; Zajonc, 1980; 1984) argued that it is a fight-flight preparation, with corresponding arousal and emotions. If one fires a pistol right next to an individual without warning, the loud noise could very well trigger an immediate, inherent alarm reaction. The alarming nature of a shot is demonstrated in Ekman et al.'s (1985) description of the common reaction of all subjects. These reactions include: lowered eyebrows, tightened eyelids, jaw muscles, body musculature, especially neck-muscles, and generally raised physiological arousal. These are essentially the same features that are observed when an individual prepares to engage in a risky situation, such as a fight (Tomkins, 1962). All subjects reported an emotional experience accompanying, or following the shot. The uniformity of the physical responses is indication for the processing of such extreme threats to survival being reflectory, immediately followed by emotional experiences.

Ekman, Friesen and Simons (1985), as well as Allen et al. (1989) therefore argued for emotional and cognitive processes to occur interchangeably before each other, or simultaneously. Emotions and cognitions can trigger each other, and can be triggered independently.

Furthermore, Isen, Daubman and Nowicki (1987) showed that affect can influence such cognitive processes as creative problem solving, while pure physiological arousal does not. Derryberry (1988) also reported the influence of emotions on cognitive processes in a series of experiments and argued on this basis that cognitive processes alone could not account for the results. Cunningham (1988) showed that induced positive mood raised the tendency to engage in strenuous activities, such as rock climbing and bicycling. Thayer (1987) studied the interaction of arousal and emotion. He reported that problems and related emotional experiences were perceived as being worse at states of low physiological arousal such as in a depressive mood, or early in the morning, as compared to being in a happy mood or in states of higher arousal, such as after a walk.

Buck (1985) argued for emotions being present at all times, sometimes even prior to cognitive evaluations. He split emotion into three types of motivational/emotional output: one involves bodily adaptation and homeostasis via the autonomous and endocrine systems, the second one refers to facial expressions and other outward expressions, and the third component relates to internal cognitive processes. The first emotional influence was evolutionary essential to secure immediate survival, because behaviors like hunger, thirst, temperature shifts, and fear, are controlled more easily when accompanied by motivating emotions, that reward the tending of those needs intrinsically.

5.2.3 Conscious and/or Unconscious Processing of Emotions

Buck's (1985) view of emotions is not consistent with cognition centered models, such as that of Lazarus (1984; 1991) or that of Schachter and Singer (1962). In their view, first an unspecified arousal is perceived, which is then cognitively interpreted as the emotion best fitting to the situation. The view of emotions eliciting cognitions, on the other hand, was already postulated by Cannon (1927; 1929; 1932) and found support in studies on people with injured spinal cords who reported emotional experiences without autonomic arousal (Chawalisz, Diener & Gallagher, 1988).

The debate still goes on, as exemplified by Lazarus's (1984; 1991) argument that emotions cannot occur without a prior cognitive comprehension and interpretation of the stimuli, so that the correct emotion can be consciously experienced. Also, Reizenzein and Schönpflug (1992) provide a historical perspective on some of the origins of the cognitivists` view on emotions having always to be the result of conscious cognitive processes.

Lazarus (1984) and Zajonc (1984), just as others who argued the emotion-cognition topic before them, provided empirical support for either point of view, so that the solution seems to be that both are correct, in certain instances. Basic emotions, such as happiness and anger, seem to be elicited immediately, and sub-consciously, and even for images that have never reached the level of consciousness (Lewicki, 1982; 1985; Lewicki & Hill, 1987), while more complex emotions are subject to a conscious cognitive interpretation, which can even alter the basic emotion initially felt (Lazarus, 1984; Sommers & Scioli, 1986). Folkman and Lazarus (1985) found that emotional experience during exams underlies cognitive evaluations, and that coping behavior can influence emotional experiences (Folkman & Lazarus, 1988). However, the studies did not rule out the possibility that emotions can also influence exam performance, or the coping processes.

In contrast to the opinions of cognitivists and behaviorists, it seems to be agreed upon in the last years that the ***"Unconscious"*** has been reborn and accepted into the vocabulary of psychologists again (see special section on the "Unconscious" in *American Psychologist*, June, 1992). In that issue, Bruner (1992), Erderlyi (1992), Greenwald (1992), Jacoby, Lindsay & Toth (1992), Kihlstrom, Barnhardt and Tatryn (1992), Lewicki, Hill and Czyzewska (1992), Loftus and Klinger (1992), as well as Merikle (1992) discussed cognitive capabilities of the human unconscious mind. The conclusion is that simple processes, automatic processes, scripts, schemata, defensive reactions and many associative processes definitely occur below the level of consciousness and some of them even outside of an attentional grasp.

Jacoby et al. (1992) treated questions of the unconscious as problems of control a human being exerts over appraisals of events, feelings, and resulting behaviors. They argued that it is not anymore a question whether the unconscious exists, but what influence it has under which circumstances.

Lewicki et al. (1992) even argued that non-conscious information processing, is far more sophisticated than conscious strategy building. The authors gave day-to-day examples for their statement that we can perform activities at an almost automatic level of processing that most of us could neither consciously analyze, explain or even understand. Such activities are for example acquiring a language as a child, learning and playing certain sports, encoding and identifying three-dimensional objects in space and sometimes only partly visible, determining the *true* intentions of a communication partner, calculating distance, size, weight, wind, angle and power of swing when hitting a golf ball, finding the right word at the right time in a business meeting and many more situations of often called "tacit knowledge" especially those, were emotions are involved.

Murphy and Zajonc (1993) and Kunst-Wilson and Zajonc (1980), reported substantial empirical data stating that subjects can differentiate verbally and emotionally on a like-dislike scale ***before*** any cognitive appraisal can take place. For example, subjects preferred meaningless symbols affectively following a very short previous presentations over ones not shown before, thereby showing affective reactions which appear independently of cognitive interpretation. Similar findings could be produced using priming techniques at and below the optimal level (Murphy & Zajonc, 1993). Mandler and Nakamura (1987) interpreted such findings in that way that the stimuli activate "underlying representations" which are then subsequently accessed. Zajonc (1984), on the other hand, argued, that before any cognitive appraisal takes place, sensory stimuli have to be processed. These could trigger predisposed emotions independently, since direct neural pathways exist, for example, connecting the retina of the eye with the limbic system in general, and specifically the hypothalamus. Zajonc (1980; 1984) argued that Lazarus (1984) stretched the concept of cognition too far by including basic sensory receptions, thus rendering the notion of "cognition" meaningless and all encompassing. Even basic information processing on a cellular level might be called cognition if Lazarus definition is followed through to the extreme (Murphy & Zajonc, 1993).

Izard`s Four Systems for Emotion Activation

Izard (1993) gave the example of a usually "happy" person being painfully hit from behind by something or somebody. The immediate pain reaction is accompanied by anger and only when the person turns around and notices that the pain is due to an accident rather than malevolence, cognitive reevaluation and reshaping of the emotional experience begins. In Izard`s example, a wheelchair had gone out of control, with the person sitting in it and thus causing the pain seeming to be in more serious trouble then the initial victim, who changed feelings of anger and rage into sympathy and sadness, although the pain may still have been there. Izard derived from this and similar examples the following stages of emotional processing:

1. *continuously active neural systems,*
2. *sensorimotor systems,*
3. *motivational systems,*
4. *cognition.*

Applied to the example above the stages are translated as follows: The basic trait emotion of the person being hit, namely a "happy mood" is due to a genetically disposed neural system, while the emotional change to those of perceived pain and the anger fight-flight response are the consequence of automatic, sensorimotor system responses and motivational systems (pain).

Finally, at the top of the hierarchy is the next emotional change towards feelings of sympathy and sadness due to cognitive information processing.

Similarly, Izard (1993, p.69) split the necessary information processing into four similar levels, namely cellular (e.g. DNA decoding), organismic (e.g. pain induced anger), biopsychological (fears connected with fight-flight situations), and cognitive. The first three groups are not dependent upon consciousness or cognition, although interactions of cognitions (e.g. memory based information) and genetically predisposed reactions do occur regularly on the higher levels.

Izard (1972; 1993) argued that developmental data from babies and infants, as well as cross-cultural research and evolutionary considerations lead to the conclusion that emotion systems preceded the cognitive ones in the evolution of human beings and outpaced them in ontogeny. Consequently, although all higher emotional systems depend on the lower ones, such as neural functioning, all systems are independently capable of eliciting higher or basic emotional experiences, such as pity or odor-disgust reactions. Izard (1993) provided empirical support for his assumptions of emotional experiences without cognitive and conscious processing. According to the literature cited, emotions can be induced by:

- *changes in levels of neurotransmitters,*
- *electrical stimulation,*
- *changes in cerebral blood temperature and subsequent neurochemical processes,*
- *visually and acoustically conditioning fear in rats, of which the visual or acoustical cortex have been ablated,*
- *sensorimotor changes, such as facial expressions, muscle action potential or body posture* (see also Stepper & Strack, 1991),
- *taste and odor.*

Also, cognitive processes may contribute to elicit emotions by:

- *motivational and "drive" systems (joy, sadness, hunger, pain),*
- *other preceding emotions,*
- *appraisals and evaluations,*
- *comparison, categorization, inferences, and decisions,*
- *attribution and belief,*
- *memory and anticipation.*

Izard (1993, p.84) summarized his four stage model of a hierarchical organization of emotion activating systems in the following way:

"Neural systems work continuously to maintain the background emotional experiences that are manifested in stable individual differences such as positive emotionality and negative emotionality. The sensorimotor systems operate in early infant-mother interactions to facilitate social communication and strengthen socio-emotional bonds. The motivational systems activate emotions whenever drive states become sufficiently intense to disturb homeostatic and autonomic processes or conditions and increase behavioral alternatives. The cognitive systems activate emotions when interoceptive or exteroceptive input requires appraisal, comparison, categorization, inference, attribution or judgement".

Leibnitz 1765 Concept of "Apperception"

To conclude the debate on conscious versus unconscious processing and emotions preceding cognitions or vice versa, I want to add another perspective on the topic combining both postulates. The issue seems to be similar to the mind-body problem, hotly discussed for centuries. Arguments had been found in favour of two independent systems and in favour of only one. It is now commonly agreed upon, that a mind without a body cannot exist in our physical world, and that a body without a mind lacks an important aspect as well. The same arguments hold for emotions and cognitions. A person without either one is unable to function and survive in our world, since emotions serve as motivators and rewards for behavior, and cognitions to design their execution. {I would even argue that it is not even possible to find such a person, since even the Vulcan "Mr. Spock" admits to having emotions, but just controls them through logic and rationality (Duane, 1988).}

Reasons for a cognition-only restriction that some cognitive psychologists take, may just be that then they do not have to deal with irrational, hard to measure and individually differing concepts, such as the unconscious and emotions, but can treat all brain functions and resulting behavior the same way, similar to the behavioristic approach of the decades before.

From a physiological point of view, however, emotions and cognitions take place in the same organ, the brain, involve the same electro-chemical processes, and even utilize some of the same structures, such as the limbic system and the neocortex. The split is thus rather hard to justify.

Furthermore, a discussion whether initial perceptions and "appraisal" reactions are cognition has probably been solved 200 years ago by the philosopher Leibnitz (1765/1949). Leibnitz stated that any individual sensory reception forms only in unity with many other so called "petite perceptions" an ***"apperceptive"*** picture, which is preconscious and only leads to conscious cognitive perception in the next step.

With the process of ***"apperception"***, used by Leibnitz (1765/1949) and later on Herbart (1816/1964), the problem of emotional appraisal, taking part in the apperceptive cognition, followed by conscious cognition, can be solved elegantly, and reduces the conflict of calling apperception cognitions or not a mere semantic one. The realization that emotions and cognitions are invariably connected and that virtually no decision (outside of the cognitivists' laboratory) can be made without emotional component, and vice versa, leads us to the more important aspects of what type of cognitive-emotional process is determined by what kind of person/situation interaction, and when, and how they are individually experienced.

5.2.4 Empirical Support for an Emotion - Risk Taking Link

Toda (1980) applied a "schema concept" to emotions and argued that emotions follow an inherent, genetically controlled experience-action response, which is dependent upon arousal. He argued that emotions have developed to lead the individual towards making the right decisions in a primitive and wild environment. Analytic abilities have developed phylogenetically later and served as a supplement to arousal-emotional automatic action patterns.

Emotions and Evolution

Toda (1980) argued that emotional schemata can be explained rationally, when one applies evolutionary theory. He interpreted them in the light of the natural situation they were originally developed for. Feelings such as anger, surprise, and fear raise the arousal level, focus attention onto the stimulus, and prepare the body for a fight-flight response, since this reaction was beneficial for survival, and is carried over into modern artificial societies. Similarly, joy is more often experienced if something unexpected happens than in response to expected events, such as an unexpected salary raise versus an expected one, or the unexpected meeting of a very good friend versus the planned visit.

The expression of emotion became essential for social interaction and for competition among species, as it represents the power, determination and status of the competitors while avoiding serious physical injury. Strong support for genetic influences in the development of emotional expressions comes from Ekman, Sorenson and Friesen (1969) and Ekman, Friesen and Simons (1985), who showed that facial expressions of emotions can be recognized independent of culture. Furthermore, Shaver, Schwartz, Kirson and O'Connor (1987) argued that the majority of people responded with similar emotions to situational descriptions, and reported similar antecedents and emotional experiences to a given situation.

One conclusion of evolutionary considerations and of genetic programming is the assumption that certain types of emotions are commonly experienced by all human beings regardless of context and culture (Smith & Ellsworth, 1985). Mauro, Sato and Tucker (1992), as well as Russell (1991) examined some of the existing literature on the effects of culture on emotion and found that many basic emotional experiences (joy, love, fear, anger) and appraisals (e.g. pleasantness, certainty, coping ability) are shared while complex emotions (e.g., Schadenfreude, Amok) and appraisals (e.g., control, anticipated effort, responsibility) did show significant differences between cultures (Mauro et al., 1992). Also, the categorization of emotions follow a similar pattern of simple categories being pan-cultural and complex ones being culture-dependent (Russel, 1991). These findings supported Mandler's (1984) argument that emotional patterns are related to situations, arousal states and cognitions in a form of schema, script, or prototype.

Also, anticipatory pleasure is reported to be higher in situations with some degree of uncertainty than in those of certain outcome. Rothbart (1973) reported that laughter generally follows the "suspense-mastery" or "arousal-safety" sequence. Children often play games that are on the edge of being just in or just out of control, thus rewarding the playful learning of fight-flight coping behavior, and the mastery of suspenseful situations. Similar patterns are reported by Piët (1987) for the enjoyment of dangerous activities, such as mountaineering, or being a stuntman. Evolutionary considerations would suggest, however, that the emotions related to surprise could also have been dangerous, as it could have led the individual to forget about potential dangers, such as other predators, who would have fought for the food found by surprise. The arousal experienced with joy, even the showing of teeth and bodily expressions, such as bouncing with joy, or the bragging of children and adults alike after receiving a valuable gift, have originally been both preparations to fight for the preservation of the joy-eliciting stimulus. The bragging behavior was interpreted as an attempt to raise one's status through this gift, which in turn increases personal control over the environment (Toda, 1980).

To develop a cognitive component of emotion is advantageous from an evolutionary standpoint, since cognitions provide the individual with an internal representation of the external world. This in turn, facilitates the anticipation and adaptive behavior of the individual, as well as the opportunity to control and analyze emotions. This ability increases self-regulatory processes, both internal ones, such as arousal, as well as their external expression. Optimally functioning self-regulatory processes could be advantageous in fight-flight situations.

If emotions are viewed as being genetically programmed survival components of human life and automatic motivators of arousal and risk taking, the necessity for a close examination of their conscious versus not-conscious, automatic processing of risk appraisals, and risk-decisions as well as risk taking behavior becomes apparent.

The Influence of Emotions on Risk Taking Behavior

Some studies related emotions and risk taking behavior directly. Johnson and Tversky (1983) could demonstrate that subjects in an induced mood of happiness perceived situations as less risky than subjects in an induced sad state. The finding suggests, that emotionality can override rational decision making in risk taking situations and should thus be strongly considered. Similar results of positive emotions facilitating risk taking tendencies and positive evaluations of risky situations have been found by Isen and Patrick (1983) for positive versus neutral mood, and by Deldin and Levin (1986) for positive, neutral and negative emotions.

Isen, Nygren and Ashby (1988) showed that the anticipated emotional impact associated with the risk of a loss is greater for persons feeling happy, since they want to preserve their state of happiness and have therefore more to lose emotionally than a person who felt bad to begin with. Combining utility and risk homeostasis theory, the findings could be explained the following way: For the person in a bad mood the expected "relative degree of gained misery" is smaller when risking a loss (less to lose), so this person has a subjectively higher discrepancy towards the optimal outcome, so the target level of risk can be set higher. Thus, the person can choose the game with the higher probability of loss. Emotions seem to have a strong influence on the behavioral compensation for perceived risk.

Finally, it was postulated that the desire to control and master risky situations is the necessary counteragent of risk taking. For example, Smith and Ellsworth (1985) found that arousal and pleasantness show up consistently in studies on emotions. They reported that emotions and situations interacted with factors of anticipated effort, pleasantness, certainty, attentional activity, self-other responsibility and control, and situational control. Thus, the relationship between risk taking and control will receive special attention in the next chapter.

Chapter 6: What Controls Risk Taking Behavior ?

So far I have discussed how emotions can motivate risk taking and that intrinsic reward systems influence this process. In chapter three, I showed that risk taking is necessary to improve chances for survival in a changing and competitive environment. It was further demonstrated that the environment imposes constant changes, and thus high levels of uncertainty on an organism. A reduction, or correct prediction of dangers would greatly increase the likelihood of coping with risks and thus the chances of survival (Paterson & Neufeld, 1987). One can achieve reduced levels of uncertainty either by gaining (primary) control over a situation, or by changing one's thoughts and desires to better fit the situation and thus gaining (secondary) control.

I shall postulate an inherent motivation to gain influence, mastery and thus control over one's environment, based on the argument that a desire to gain personal control is the necessary and advantageous counteragent to risk taking from an evolutionary standpoint. Thus, it is also most likely genetically controlled and intrinsically reinforced, and can be result, cause, motivator, and inhibitor of risk taking.

I shall first define concepts of control, before presenting support for the assumption that personal control and mastery are emotionally and physiologically intrinsically rewarded. Implications for clinical and vocational interventions, *Locus of Control* and the *Two-Process Model of Perceived Control* will be presented.

Studies showing that even an illusion of control can lead to better medical and psychological well-being will further strengthen the importance of the concept of control. Finally, the relation between control and risk taking will be discussed.

6.1 Personal Control: Definitions and Historical Concepts

Thompson (1981, p.89) defined control as:

> *"the belief that one has at one's disposal a response that can influence the aversiveness of an event".*

However, control can also be expected and gained over positive outcomes (Langer, 1975), and so I expand the definition to the more accurate one:

Personal Control is the belief that one has at one's disposal a response that can influence the outcome of an event.

This definition can incorporate Rothbaum, Weisz and Snyder's (1982) distinction between primary and secondary control. Rothbaum et al. (1982) disputed that submissiveness, withdrawal, and passivity, have to be signs of relinquished personal control. They argued, that in addition to bringing the environment in line with one's wishes (primary control), one can also bring one's wishes in line with the environment (secondary control). In the case of secondary control, the influence over outcomes is gained by withdrawal, passivity and interpretative change of one's wishes or the perception of the situation ("Always look on the bright side of life"). Instead of engaging in behavior, in the case of uncontrollable outcomes, secondary, passive control mechanisms may be the only way to cope with the situation and to retain some control over one's future (Rothbaum et al., 1982). Both types of control serve the basic need to gain influence over the environment to increase one's chances for survival.

A basic need for control has often been postulated. Adler (1930) proposed that an individual's major motivational force is to demonstrate one's competence and mastery over events. Fromm (1941) discussed the "process of individuation" with its key components of increased freedom, loneliness and powerlessness. Fromm (1941) stated that individuals often flee from this process of individuation to regain the sense of power through association with more powerful others. The position of power, which was represented during childhood by their parents, is now substituted by identification with, and submission to other powers, such as religion, capitalism, fascism, or compulsive conformism. This process of gaining power and control through something outside of the person, compensates for the feeling of powerlessness the individual's self experiences (Fromm, 1941). Control is thereby gained differently by different people, but works as a strong motivator for behavior, emotion and cognition.

Hendrick (1943) talked about an instinct to master, and Kelly (1955) described individuals as constantly comparing and adjusting their expectations about outcomes against their perceptions about the present situation, in an effort to gain optimum predictability and control of the future. White (1959) talked about a need for competence, and DeCharms (1968) posited an inherent *"effectance motivation"*. In his theory the reward for interacting with the environment is the feeling of competence one obtains from exerting control. DeCharms (1968) assumed that due to different life experiences, individual differences influence emotional experience of mastery. DeCharms (1968, p.269) stated:

"man's primary motivational propensity is to be effective in producing changes in his environment. Man strives to be a causal agent, to be the primary locus of causation for, or the origin of, his behavior; he strives for personal causation."

Kelley (1971) suggested that we apply causal analyses and attributions to explain events as a means of effective exercise of control over our world. He argued that we attribute to controllable sources those events over which we ***desire*** the greatest amount of control, whether the influence is real, or an illusion of control. A connection between arousal caused by suspense and feelings of control or mastery was found by Rothbart (1973). Rothbart (1973) suggested that danger-seeking may be rewarded by feelings of mastery and competence if one manages successfully a dangerous, and suspenseful situation. The higher the stakes, the larger the mastery, the more pleasant the emotional reward. Weiner (1985) argued that success, when occurring in response to a challenge implies a measure of effort, thus can be attributed to one's own credit.

Deci and Ryan (1987) split intentional behavior into self-determined or autonomous, and controlled behavior. In contrast to control as personal influence (Rotter, 1966), Deci (1975) used "control" in the meaning of: "influenced from the outside". Deci and Ryan (1985) stated that controlled intentional behavior is invariably determined by conditioned reactions to intrapsychic or environmental factors and is therefore not representative of true choice. Autonomous behavior, on the other hand is, according to Deci and Ryan (1987, p.1025): *"an organismic concept anchored in the sense of a fuller, more integrated functioning"*.

Self-determined versus controlled are considered opposite ends on a determination dimension. The authors argued that only self-determined behavior increases intrinsic motivation, whereas controlled behavior decreases it. This leads to the conclusion, that only voluntary risk taking behavior is intrinsically motivated and self-perpetuating, while involuntary risk taking behavior is discouraged. Deci and Ryan (1987) stated that more self-determined people regard potentially dangerous situations as a challenge, while more "controlled" people experience the same situations as threats.

Secondary control relies on evolutionary young, complex cognitive interpretations, while primary control may work on lower cognitive levels. I now want to focus on development and manifestations of primary control.

6.2 Personal Control and Evolutionary Utility

In chapter three, I suggested that risk taking behavior is performed at a subjectively optimal, intermediate risk level, as this optimizes the likelihood of an organism and its off-spring to survive. The necessity to be curious and exploratory in an uncertain and changing environment was demonstrated. However, the desire to satisfy one's curiosity, and the desire for security can be mutually exclusive, although both are important factors to fulfill the need for survival (Berlyne, 1960; 1971; Solomon, 1980).

Curiosity, the necessity to provide food and water, and the dangers of changing environments forced organisms into situations with considerable uncertainty, while the motivation to be secure rewards reducing risks to a minimum (Hull, 1943). One mechanism to optimize this relation of opposing needs is to gain personal control. Averill (1973, p.290) stated:

"An animal (especially an unspecified primate of the type ancestral to man) who did not seek information about, or attempt to exert control over, potentially harmful events probably did not survive to contribute to the evolution of the species. This would lead to a biological predisposition for personal control, even though under certain conditions the exercise of that control might be stress inducing rather than stress reducing. Of course, similar reasoning could be applied to ontogenetic as well as phylogenetic development. The growing child is taught to assume responsibility and exercise control, even though he may at times find this rather frightening. Once socialized, the preference for control may be manifested regardless of whether or not such control is actually effective in reducing stress in any particular instance. In other words, the desire for personal control may be a deep-seated motivational variable, whether phylogenetically or ontogenetically based."

Given the assumption that different species and members of species have developed different optimal levels of risk taking one can apply similar reasoning to control. Consequently, organisms experiencing major changes in their environment and high levels of uncertainty, should develop a stronger need for personal control than organisms experiencing a relatively stable environment.

Furthermore, if successful in surviving under changing and uncertain conditions, one could even expect a high ability of these individuals to influence the environment. In addition, one could assume that the individuals with a higher desire and ability to exert personal control also have a strong belief in their personal control. To acquire personal control, therefore, serves as a measure to reduce uncertainty, and the belief in personal control alone, should be able to have the same effect of reducing the level of anxiety, regardless of the correctness of this perception.

The importance of perceived control has also been shown in other approaches to predict behaviors from attitudes. While traditionally weak correlations were found, Fishbein and Ajzen`s (1980) approach to use attitudes to specific behaviors was of greater success. Later research on the theory and its modification into the theory of planned behavior (Ajzen & Madden, 1985; Doll & Ajzen, 1992) showed, however, that the original factors had to be completed by another one, namely perceived control.

The necessity to gain control over arousal, stress and anxiety was further demonstrated by Frankenhäuser and Johansson (1986), who showed that continuous high levels of stress and arousal lead to serious internal damage. The literature on "burnout syndromes" can offer further indications as to the long-term negative consequences of constant activation of the fight-flight energy reserves.

6.3 Personal Control and Psycho-Biological Consequences

Clinical consequences of perceived personal control, or the lack of it, are clearly demonstrated by the "learned helplessness syndrome" (Seligman, 1975). Seligman (1975), Hiroto and Seligman (1975) and Miller and Seligman (1975) showed the debilitating effects of prolonged exposure to uncontrollable aversive stimuli. Depression and poor performance resulted from this helplessness training, while subjects who perceived some control over events did not display these helplessness effects.

Apart from the automatic reduction of arousal through an opponent process, or reinterpretation (chapter three), the development of voluntary and cognitively controllable mechanisms would also have advantages for efficient energy-housekeeping, and for survival. One such voluntary cognitive concept is that of *Locus of Control.* Heider's (1958) concept of internal versus external *locus of causality* attributed intrinsically motivated behavior to an internal locus of causality and vice versa. This split was the beginning of research on internal versus external locus of control, which was advanced by Rotter (1966).

The belief in personal control - or the lack of it - is extensively dealt with in Rotter's (1966) Locus of Control Scale. People with an internal locus of control perceive that they control the probability of gaining positive reinforcements, while people with an external locus of control attribute the outcome, in any given situation, as being controlled by others, chance or fate (Rotter, 1966). Bandura (1977) added that expectations of competence are also necessary for internal control. The research conducted on the theory in general was reviewed by Strickland (1990) and Rotter (1990).

Marshall (1991) examined the effects of different dimensions of an internal health locus of control. He found that a multidimensional construct emerged with self-mastery being strongly related with indexes of physical health and well-being. The findings suggested that believing in one's own abilities in changing the health status may actually produce a significant influence. The authors claim that it is unlikely that by answering the internal locus of control items the subsequent answers on 33 current health problems (e.g. high blood pressure) will be changed.

Thompson et al. (1993) showed that cancer patients who believed to have some control over the outcome of the disease were significantly less depressed than those who were not and were much better able to deal with daily hassles and problems. They suggested that these beliefs, although possibly irrational, may very well be beneficial to the course of the disease by providing the patient with more energy, resilience and dedication, as well as by improving communication with family and medical personnel.

Also, Emmons (1992) showed that high-level striving with high levels of perceived control did lead to significantly fewer physical illnesses and to a significantly better psychological well-being. Thus, perceived control seems to influence people throughout their psycho-medical health status.

Lefcourt's (1973) review of studies dealing with aversive stimuli demonstrated that perceived control reduced physiological as well as psychological stress reactions. More support for the detrimental effects of too low control can be found in studies on predictable versus unpredictable shock and stress. Under both conditions, stress was experienced as higher when subjects had no control over the delivery of the shock (Arthur, 1986; Abbott and Badia, 1986). Melzack and Casey (1970) suggested the experience of pain to be dependant upon the person's interpretation of pain. The importance of personal control in situations of danger or pain was supported in their *"Gate-Control Theory"*. The authors postulated a psycho-physiological, situation-dependent "gate" that regulates the experience of pain. Thompson (1981) reviewed the literature on perceived control and the perceived aversiveness of stimuli. She split control into a behavioral, cognitive, informational, and retrospective component and showed that each component reduced the perceived aversiveness of stimuli successfully.

Ryan and Grolnick (1986) reported that children who felt externally controlled were more aggressive than children who were in self-determination supporting classrooms. Langer and Rodin (1976) studied institutionalized elderly and found significant improvements in physiological, medical and psychological evaluations, after the inhabitants had been involved in interventions strengthening self-determination, personal control and autonomy.

King (1984) found that people scoring high on perceived autonomy regarded cardiac surgery more as a challenge than as a threat, in contrast to those subjects scoring low in self-determination and autonomy. Frankenhäuser et al. (1980) reported a relationship between the sympathetic-adrenal and pituary-adrenal response to achievement situations, and the Type A/B personality pattern.

These studies further indicated the important role perceived control plays in moderating psycho-physiological and medical correlates of well-being or stress (Weigel, Wertlieb & Feldstein, 1989). Those subjects who experienced more personal control, also reported better physical, emotional and cognitive health.

These findings strengthen the postulated link between one's inherent desire for controlling one's environment and mechanisms intrinsically rewarding actions directed to increase the level of control. Rewards, such as emotional well-being, increased activity and improved physical health, are functions that are advantageous from an evolutionary standpoint, are motivating, and self-perpetuating, as they increase personal control.

Berlyne (1960, 1971) and Solomon (1980) stated that both the tendency to take risks and the desire to gain personal control must be rewarded by an organismic feedback, either through the limbic system, or through adreno-cortico arousal systems. Frankenhäuser and Johansson (1986) studied relations between high-control and low-control situations on one hand, and self-reported, and physiologically measured stress on the other hand. They found that under conditions of low control perceived stress is significantly higher and the excretion of cortisol is also significantly increased. After controlling for physical effort, adrenaline was elevated in both high and low control situations, due to automatic adrenaline release in response to heightened arousal. Increased adrenaline levels were always related to arousal and stress. In passive, helpless states, with feelings of distress and low personal control being predominant, increased cortisol secretion was found, whereas in the active, high-control situation decreases in cortisol were observed. These findings supported their effort-stress model, stating that adrenaline is secreted to any kind of active, controllable effort, while the perception of stress is a function of perceived controllability, which in turn determines whether cortisol will be elicited.

6.3.1 Personal Control, Stress and Vocational Behavior

Frankenhäuser and Johansson (1986) found further support for their effort-stress model in monotonous, low personal control jobs, like work on assembly lines (Johansson et al., 1978) or operating computers (Frankenhäuser & Johansson, 1986). In both cases they found elevated cortisol secretions and reported levels of stress, compared to control groups holding similar jobs, yet having high personal control.

In a longitudinal study, Johansson and Aronsson (1984) found that heightened levels of perceived stress and adrenaline excretion persisted even hours after the job was finished and the subjects were at home, relaxing. The persisting negative effects of passive, monotonous, low-control work, even to times when the work is done, was also reported by Rissler (1977), and Frankenhäuser (1981) for a four week period of overtime, during which adrenaline levels were constantly elevated, both on and off the job. Frankenhäuser and Johansson (1986, p.297) stated:

"we have emphasized work conditions which stimulate ability to exert control and influence... In the short term, they increase job satisfaction and well-being and hence reduce neuroendocrine stress responses. In the long term, being able to exert control over one's work may help in developing active ways of coping with both minor and major stressful life events."

Enzle and Anderson (1993) found that according to the level of perceived control, and the perception of the control oriented intentions of surveillants in an experiment, subject's intrinsic motivation was influenced. When the surveillants were perceived as control-oriented the motivation dropped.

Aspinwall and Taylor (1992) found in a two-year longitudinal study on college students that self-esteem and control accounted for high levels of motivation, and higher grades, even after controlling for college entrance exam scores. They interpreted the data in support of Taylor and Brown`s (1988) *Cognitive Adaptation Theory*, although the results have been mediated by better coping strategies.

Schmitz and Skinner (1993) found similar results for high school children. In both cases coping strategies and perceived control seem to be linked with motivation, effort and success at academic performances. Furthermore, studies by Dabbs et al. (1990) indicated that hormonal levels of testosterone may have a subtle influence on occupational choice.

6.4 Perceived Control and Risk Taking Behavior

Piët (1987) assessed stuntmen's motivation for engaging in that risky profession. She found that many subjects did not perceive risks as inducing fear, but more as a chance to improve their mastery and control over their lives. According to Frijda (1986) and Smith and Ellsworth (1985) challenge is connected with higher control and less uncertainty than fear. Piët (1987) stressed that stuntmen are very convinced of having high personal control in their profession, yet they feel more uncertain about outside actions, such as driving in normal traffic. Piët (1987) suggested that stuntmen perceive the objectively risky actions in their profession as a challenge. According to Piët (1987, p.210) stuntmen have a:

"more or less generalized sense of the controllability of events, or upon a tendency in the subjects to make it be or appear controllable... The reward of danger seeking is thus to be found in the sense of mastery and competence"

Furthermore, Piët (1987) argued, that the strong exposure to dangerous events subjectively and objectively increases the abilities of the stuntmen to control such situations. The ability to stand the strains of confrontation with grave risks and to concentrate under highly stressful situations leaves them with a rewarding sense of mastery and "feeling alive".

Research related to risk taking found some observational support for the influence of perceived personal control on behavior in comparing locus of control, risk taking propensities and entrepreneurship (Ahmed, 1985). This author reported positive correlations between risk taking, having an internal locus of control, need for achievement (McClelland, 1961) and being an entrepreneur. In another study, the relationship between involvement in fatal driving accidents and previously assessed measures of internality and externality of control was explored (Montag & Comrey, 1987). After adapting the Rotter (1966) "locus of control scale" towards driving Montag and Comrey (1987) found driving-internality to be negatively related, and driving externality to be positively related to involvement in fatal accidents. Wuebker (1986) also found correlations of a modified safety locus of control scale and industrial accidents, as well as attitudes towards safety. Bragg and Finn (1982) reported that subjects driving as a passenger expressed greater perception of risk, due to a lack of control over the likelihood of an accident. People who attribute accidents and events externally seem to be less careful about health, driving and their jobs than people attributing such events internally. Newman and Benassi (1989) reported that the context of a situation significantly influenced the judgement of control over experimental situations with varying degrees of uncertainty.

In experimental research on the locus of control and financial risk taking, Strickland et al. (1966) found that under certain conditions internally controlled subjects bet more money in a dice game than externally controlled ones. When subjects were allowed to place a bet *before* throwing the dice, internally controlled subjects bet significantly more money than when the bet was placed *after* throwing the dice, but just prior to discovering the result of their throw. Externally controlled subjects showed no significant difference in their respective bets.

Strickland et al. (1966) explained their findings about internally controlled subjects betting more money before a throw as due to a "superstitious" belief they could influence the outcome of the throw. When the dice had already been rolled, the *parapsychological* possibility of "telekinesis" was reduced, thus the hope for a win, and the perceived control smaller, which made them bet less money (Strickland et al., 1966).

6.5 Illusion of Control

Langer (1975) found an interaction of illusion of control and competition, familiarity, choice and involvement, even in situations governed clearly by chance, such as a lottery. Burger and Cooper (1979) confirmed Strickland et al's (1966) findings and showed that subjects scoring high on the *Desire of Control Scale* (Burger & Cooper, 1979) fell more victim to this illusion of control in a game of pure chance than subjects scoring low on the scale. Burger (1987) showed that greater choice, also perceived as greater control, lead to better performance in a learning task. Burger (1986) also found that just being familiar with a coin tossing task led to higher perceived control and higher bets, than when subjects did not know the game.

Langer and Roth (1975) showed that students participating in a rigged coin tossing task showed a greater illusion of control, as indicated through self-report and prediction for upcoming trials, if they were rewarded for supposedly good performance in the beginning of the task, which gave them a feeling of mastery over the game. When subjects failed on the first trials, later success did not provide the feeling of increased control and mastery. This experiment revealed that independently from being high and low in desire for control, or being an internally or externally controlled person, feelings of personal control, and the amount of economic risk an individual takes can be greatly influenced by situations.

Burger and Schnerring (1982) and Burger and Smith (1985) studied gambling behavior and found that the desire for control scores predicted the type of game that problem gamblers bet upon, namely those with a hint of controllability. The authors also found a relation between desire for control scores and highest amount of financial loss during a gambling year, thereby validating their laboratory findings in a real-life sample and situation. This finding may explain why gambling can become so hard to stop. If gamblers believe to be in control over the game, than they can afford to lose some games, because in the long run their skill will eventually let them win their money back. Henslin (1967) studied dice playing and found that players thought a soft toss produces low numbers and a hard toss high ones. The players believed that effort and concentration would pay off.

Taylor and Brown (1988) reported that people who have a strong tendency to perceive the world overly positive, who overestimate their personal control and who maintain a positive illusion also tend to be more happy, successful at work and in social interactions. People with accurate perceptions of control and the environment, however, tended to be more depressive and less contempt with their lives. Gollwitzer and Kinney (1989) further specified the benefits and drawbacks of illusion of control. They found that overly optimistic perceptions increased goal striving, while accurate perceptions tended to increase "sound" decision making. An interesting finding regarding possible undesired effects of high personal control is reported by Burger (1989). When control is defined as:
"the perceived ability to significantly alter events" (Burger, 1989, p.246), situations can occur, where personal control is seen as less desirable. This low desirability of control occurs when:

a) it leads to an uncomfortable level of concern for self-presentation, or
b) it decreases the likelihood of success, or
c) it leads to increased attention on the negative aspects of the situation.

Such conditions may lead a person to either forfeit the chance for personal control, or lead to unpleasant feelings related to perceived control. I argue, however, that these conditions are in reality a lack of personal control over the outcome. If someone is given the "control" of deciding to either fly a jet-plane oneself, or to leave that job under the control of an experienced pilot, most non-pilots would forfeit their personal "control" over the airplane in order to increase the control over a safe arrival.

6.6 The Two-Process Model of Perceived Control

In the example of flying in an intercontinental jet, the passivity of just sitting in the back, or the withdrawal from steering oneself, or even the submissiveness to the authority of the pilot could be understood as examples of relinquishing one's control over one's future, since these behavior patterns are identical to the *"helplessness"* behavior described by Seligman (1975) or Lefcourt (1973). Rothbaum et al. (1982) would argue, however, that our likelihood of survival and our future are much better taken care of by not taking primary control. Although the outcome may be equally predictive, the need for survival would urge us to remain passive. The theory of cognitive dissonance reduction (Festinger, 1957) would also predict that if we perceive a discrepancy between what we want and what we perceive, we either change the things we perceive (primary control) or the things we want (secondary control) by cognitive reevaluation.

Rothbaum et al. (1982), and Weisz and Stipek (1982) split secondary, or behaviorally passive control into four groups:

1) Predictive control refers to the effect that an attribution to one's own limited abilities or to an aversive event can serve as a predictor of the future and therefore as a means to control disappointments. Empirical support for this type of control was found by Bandura and Wood (1989), who showed the relation between perceived control and satisfaction with performance standards. More support was provided by Mettee (1971) who found that subjects were actively striving to avoid disappointments, even when that meant to avoid some form of success, if that success increased the probability of eventual total failure.

2) Illusory control occurs in situations where the outcome is attributed to chance and the person perceives his/her strength in being lucky. Em-

pirical support for this type of control is found in the common belief that praying, reading the horoscope, and other forms of superstitions have an effect. Experimental support is given by Langer (1975; 1977) who showed that gamblers hope for the "lucky streak" in games of pure chance.

3) Vicarious control refers to identification with and attributions to powerful others, thereby gaining a substituted control. Impressive examples for this type of control can be found in group membership, such as the Ku-Klux-Klan, or other Neo-Fascist groups. The pilot example, or a child's identification with the parents, (demonstrated in experiments by Bandura, Ross & Ross, 1963) are further examples of vicarious control. Fromm (1941) also regarded the belief in God, the nation, the conscience, or a psychic compulsion, such as conformism as forms of secondary control.

4) Interpretative control refers to the observation that often people try to understand and derive meaning from otherwise uncontrollable events, such as severe injury or death, in order to be able to accept them. Studies by Bulman and Wortman (1977) revealed that paralyzed accident victims exerted considerable effort to make sense of what has happened to them. All subjects tested quarrelled with the question: "Why Me?" and had an explanation, such as fate, God, chance, or deservedness. Also, the victims reevaluated their injury in a more positive light, such as: slowing down, or learning about other people and life. Lazarus and Launier (1978) regarded these positive thoughts as the aim of cognitive reinterpretation, while Bulman and Wortman (1977), and Rothbaum et al. (1982) saw the interpretative control itself as rewarding action, providing intrinsic rewards through perceived control over the future.

To gain either primary control, or if that is not possible secondary control over past, present or future events may also serve the purpose of staying mentally healthy. The randomness of fate, the inevitability of death and the sometimes irrational and incomprehensible events and accidents are very threatening for organisms whose survival relies on being able to perceive, predict, and explain causal relations in order to optimize the relation between necessary risk taking and dangerous uncertainty.

6.6.1 Desire for Personal Control and Risk Taking

The above cited research has described in numerous ways how the perception of personal control can influence behavior. Instances were reported where risk taking in gambling (Strickland et al. 1966; Burger, 1986), driving (Montag & Comrey, 1987), and engaging in dangerous professions (Piët, 1987) was influenced by a desire for personal control or belief in personal control. The findings also suggested that greater predictability is associated with less arousal (Carr, 1984; Katz & Wykes, 1985), and that control allows a person to alter the threat-response directly, or through a behavior altering the source of the stress (Paterson & Neufeld, 1987).

Conflicting information exists, whether internal locus of control leads to increased or reduced risk inclination (Ahmed, 1985; Montag & Comrey, 1987). Inconsistencies are also found in scales of: desire for control, locus of control, and the autonomy-controlled dimension (Burger & Cooper, 1979; Thompson, 1981; Deci & Ryan, 1985). Common factors are:

a) the assumption of a need for personal control and mastery;
b) an intrinsic reward system for achieving control and mastery;
c) relatively large individual differences in belief, attribution, perception, and experience of personal control; and
d) a connection between behavior under conditions of uncertainty and the experience of personal control.

With the desire for risk taking counterbalanced with the desire for mastery and control, people seem to be involved in a constant readjustment and adaptation process. Different environments and situations trigger risk taking behavior. Boring circumstances produce challenge seeking behavior while dangerous situations lead to increased efforts to control for negative consequences and unwanted results.

People seem to engage in risk optimization behavior to achieve a level of relative excitement and security that befits their desires for any given moment. Inadequate value systems, errors in risk perception and risk estimation or inabilities in proper behavioral alternatives, as well as a lack of training lead to dangerous consequences, accidents and possibly death. Furthermore, cultural norms and values promote more or less risky behavior and may "seduce" susceptible people, especially young men, to engage in behavior beyond the optimal level of risk taking.

As a summary of the chapters above it can be stated that a combination of personality and situation variables, as well as the aspect of utilizing utility approaches in day to day behavior, and the influence of actual and perceived -even illusionary- control significantly influence the motivation for risk taking behavior. Apart from the internal changes risk perceptions, emotions and cognitions cause on attitudes and judgements, there are also consequences of the risk taking behavior in the outside world. Those consequences of risk taking behavior may be positively experienced, if the desired goal was reached or extremely negatively, as in accidents. Thus, consequences or risk taking behavior will be the focus of the next chapter.

Chapter 7: What Are the Consequences of Risk Taking Behavior ?

The consequences of risk taking behavior can be manifold. It can lead to financial gains, social fame and praise, the desired mating partner and many other positive outcomes. The major concern of public and private people alike, is, however, what the possible negative consequences may be. Loss of fame and fortune are certainly some of them, but most important is the loss of health and ultimately life. The major scope of this chapter is on the loss of health and life due to accidents and unwanted events.

Accidents, especially in traffic, are the single most common cause of death in people age 16-24, and a leading cause of death for people in general. During the 1960's about 1.5 million people were killed in traffic accidents in Europe and the USA (Sleet, 1984; Statistics Canada, 1983; Svenson, 1978). In 1987, 47093 people were killed in the USA, resulting in 1.4 million person years lost (Elander, West and French, 1993). Injuries stayed at the same level of about 1.8 million - 2.0 million injuries per year in the USA (Lamm, Choueiri & Kloeckner, 1985). Thus, when participating in traffic as a driver, passenger, or pedestrian, one tolerates a considerable risk of dying. Similar arguments can be made for the approximately equal number of accidents per year at the workplace (Murphy et al., 1986). Smith et al. (1992) reported that about 3.5 million athletic and recreational sport injuries occur annually in the USA alone. The type of risk taking behavior, however, varies greatly. Some dread the risk of accidents and try to minimize personal risk by living cautiously, while others enjoy the arousal that accompanies living "on the edge and beyond".

A combination of personality and situation variables is held responsible for most traffic, industrial, and sport accidents (Scholz et al., 1983; Hale &

Glendon, 1987). I decided to look at risk taking behavior and its relation to accidents to find out whether the concept of risk and risk reduction are useful tools to reduce this incredibly high number of physical and emotional injuries. Also, by studying why people engage in dangerous activities, I hope to gain a deeper insight into the concept of risk taking behavior.

I shall review risk taking and its possible positive and negative consequences in relation to sport, industrial, and traffic accidents. Some suggestions will be made as to assessment of human reliability and safety improvements based on statistical and motivational techniques. I shall present some of the more prominent models attempting to explain risk taking in driving, such as: the accident proneness model, Näätänen and Summala's (1974) *Zero-Risk Model*, Fuller's (1982) *Threat Avoidance Model*, Klebelsberg's (1982) *Subjective-Objective Risk Model*, and Van der Molen and Bötticher's (1988) *Hierarchical Model of Risk*. Many psychological phenomena related to accidents have been explained by Wilde's (1978) *Risk homeostasis Theory (RHT)*. The relationship of behavioral adaptation and technical versus psychological safety measures will be discussed at the end. The first focus of the chapter, however, will be the overproportionally high involvement of young males in accidents

I shall argue in this chapter that underlying risk taking behavior is a homeostatic process, which attempts to compensate for deviations from the subjectively optimal level of risk taking, through corrective action. The optimal level, or target level of risk is determined by physiological, emotional and cognitive risk assessment and risk acceptance processes, combined with expectancies about their future benefit. I shall argue that accidents often result from an improper setting of the target level of risk for the given situation, from over- underestimation of personal abilities, or from faulty (wrong, or incomplete) risk assessments and risk acceptances, in other words: conscious and non-conscious ***errors***.

7.1 Young People, Risk Taking Behavior and Accidents

The young ... are full of passion, which excludes fear;
and of hope, which inspires confidence. (Aristotle, Rhetoric Book II)

Quadrel, Fischhoff and Davis (1993) used this quote to demonstrate the long held belief that young people are much higher risk takers than older ones. There has been a large body of literature showing that young people, especially males, score higher on risk taking questionnaires (Zuckerman, 1979; 1991), perceive lower risks for themselves in identical situations (Tränkle et al., 1989), engage more in risk taking activities, and finally, are more involved in accidents (Jonah, 1986).

Furthermore, based on sport activity studies, traffic fatality data, as well as on sensation seeking indices, it has to be concluded that young people, especially males between the age of 16-24 are at the highest risk, and choose to take the highest risks. For example, young people constitute about 17% of Canada's population (Statistics Canada, 1983), but they account for 31% of all fatalities and 33% of all injuries in traffic (Transport Canada, 1984). In Ontario, 16.2% of all licensed drivers were between 16 and 24, yet they accounted for 30% of traffic fatalities (Ontario Road Safety Annual Report 1987). A number of studies tried to explain this disproportionately high rate (Valentine, Williams & Young, 1978; Hodgdon, Bragg & Finn, 1982; Cameron, 1982; 1983; Jonah, 1986). The finding that especially young males are overrepresented in these data, plus the fact that young males are normally found to be high risk seekers indicates a possible causal link between risk taking and accidents.

Some studies attributed these observations to increased hormonal activity in adolescents, especially the hormone testosterone in young males. Buchanan, Eccles and Becker (1992) reviewed many of the studies on this topic in a meta-analysis. They provided a very differentiated picture of the effects hormones have on mood, thought, behavior and emotions of both adolescents and adults. They concluded that both hormonal, as well as non biological factors have a major influence.

Several studies have specifically focussed upon either risk personality factors, or risk perception and risk utilities (Eysenck, 1967; Zuckerman, 1979; Slovic, 1978; Slovic & Fischhoff, 1982; Hodgdon et al., 1981; Wilde, 1976; 1982; 1986; Wilde & Murdoch, 1982). Young drivers often had less experience (Brown, 1982), overestimated their abilities (Kogan & Wallach, 1964), while underestimating the dangers (Finn & Bragg, 1986).

Brown and Copeman (1975) reported that young male drivers rated traffic offenses as less serious than other traffic participants, and they also rated their own responsibility as less serious than other traffic participants did. Brown and Groeger (1988) and Wright (1984) also reported that young males in general and particularly young drivers overestimate their abilities. The "risk behavior syndrome" (Jonah, 1986) denotes that some young people in general engage in risky behavior (Zylman, 1972; 1974; 1975).

On the other hand, Quadrel et al. (1993) examined the hypothesis that young people are less informed about possible risks and overestimate their invulnerability. They did find, however, that adults, adolescents and high risk adolescents all overestimated their invulnerability, but that there were no significant differences between these groups. The authors concluded, that schooling, informing and advising adolescents about possible bad consequences alone will not change their risk taking tendencies.

7.2 Risk Taking Behavior and Sport Accidents

Svebak and Kerr (1989), Kerr and Svebak (1989), and Kerr and Cox (1990) found that participation in high risk sports correlated with different personality patterns, such as being more impulsive and more paratelic oriented. However, to better understand sport risk taking behavior it is important to consider the danger inherent in physical sports (Zuckerman, 1979), as they have higher accident rates than other leisure time activities, such as playing chess, or reading. Halpern et al. (1987) found that one in every 14 teenagers in Massachusetts emergency rooms was there because of sports injuries, as compared to 1 in 50 due to traffic accidents. Seventeen percent of sport injuries resulted in hospitalization compared to 14% of the traffic injuries.

Clarke (1966) calculated the average annual fatality rate per 100.000 participants and reported a rate of 3.9 for football, for power boating 16.7, for auto racing 120.0, for horse racing 133.0, and for motorcycling 278.6. Motorcycling has more than 3 times the probability of fatal accidents per hour of exposure than all other accidents combined, and two times as high a rate as all other causes of death, in the 16-24 age group.

De Haven and Lintner (1986) reported that the University of Rochester treated about 700 cases per year at the sports medicine center. Football accounted for 63.9%, followed by Basketball (5%), Soccer (3.2%), Track (2.7%), and Baseball (2.3%). He also found that 80% of the injuries were reported by males, and the peak age of the injured was from 16-19 years. US-Football is a sport with a high frequency of severe spine injuries and death (Clarke, 1966; Torg et al, 1979). Clarke (1966) calculated for 850,000 football players in high schools during the 1964 football season a fatality ratio of 1:1,2 when compared to all other accidents combined, and a fatality ratio of 1:9, when compared to traffic accidents.

Torg et al. (1979) reported fatality ratios to have slightly improved since Clarke's (1966) study. The rate of severe spinal cord injuries, however, has risen by a factor of three for all spinal cord injuries, and by a factor of two specifically for cervical quadriplegia as a result. Torg et al. (1979) attributed this rise in spinal cord injuries to the use of improved helmets and helmet face-masks, which prevent head and face injuries in a collision. These face masks allow the defenders to use the helmet as the primary point of contact in blocking ("battering ram"), tackling and head butting. The assumption is supported by the finding that "spearing" head-first into the opponent is the manoeuvre leading to the injury, and by the finding that defense players, who have a high frequency of "spearing" action also have a high rate of such injuries (Torg et al., 1979).

This behavioral and injury shift in relation to improved safety devices is a clear example for the occurrence of risk compensation processes, meaning that safety improvements are transformed into performance improvement while a relatively constant total level of risk is maintained. Since face and head are not in danger anymore, the individual uses the head to accomplish a better tackling performance and thereby shifts the type of injury risk from that of a broken skull to that of a spinal cord injury. To assess whether the total amount of risk stayed truly constant one would have to evaluate the subjective value of a death versus a cervical quadriplegia injury.

In relation to downhill skiing, Johnson et al. (1980) reported that in the 1970's about 10 million people skied in the USA. The annual injury estimates range from 165,000 - 600,000. With the great numbers of injuries that are neither reported at the ski-resort, nor anywhere else, the estimates of injuries per 1000 skier days range from 3-6, or even much higher. Johnson et al. (1980) also reviewed the literature on the effect of safety gear and concluded, that the type of injury has changed with improving safety gear. While the rate of ankle injuries dropped considerably, knee injuries, upper extremity fractures, and thumb injuries have increased, so that the total reduction in injuries is only slightly noticeable. Blitzer et al. (1984) reported children to have the same likelihood of having a ski accident as adults, and that experience is negatively correlated with injuries. Equipment adjustments by professionals did not correlate with injuries.

Bicycling is a common mode of transportation and recreation for people in the age of 17-66 (Kruse & McBeath, 1980; Kiburz et al., 1986). Kruse and McBeath (1980) reported that 13% of the bicycle population of the University of Wisconsin had an accident in the year before, and 29% in the three years preceding the study. In 62% of the cases someone had been injured, but only 8% of all accidents had been reported to the police. Kiburz et al. (1986) reported similar findings from 492 bicycle club members. Kiburz et al. (1986) also reported that every year an increasing number of both bicyclists and accidents appear, with 500,000 reported emergency room visits in 1984. Kiburz estimated the accident frequency at 31 per 100,000 km travelled for adults, and 43 per 100,000 km for children.

Petras and Hoffman (1983) examined 90,000 parachute jumps in one year in an armed forces facility. 615 injuries (0.68%) and no fatalities were reported. They concluded that parachuting qualifies as relatively safe compared to occupational accidents, while as a sport it is more dangerous.

On the other hand, Smith et al. (1992) reported that athletes high in sensation seeking had significantly less negative sport specific life events and less injury time-loss than athletes low in sensation seeking even after controlling for exposure. Smith et al. (1992) explained this finding through the attribution of coping skills. Low sensation seekers attribute feelings of anxiety to a lack of control and less coping skills. Sensation seeking may thus serve as a stress resiliency factor in sport activities, rather than supporting increased participation in activities with a high risk of injury.

7.3 Risk Taking Behavior and Industrial Accidents

Industrial accidents are the second leading cause of death - after traffic accidents - for people in the age of 18-65. Hoyos and Zimolong (1988) stated that in the USA an occupational accident which results in injuries occurs every 17 seconds and a fatal one every 46 minutes. The total costs of this accident toll is estimated to be $ 33.000.000.000 (National Safety Council, 1985). These costs can be further split into about 50 % direct costs (wage loss, medical costs, insurance costs, etc.) and 50 % indirect costs (time loss of co-workers, training of replacements, time for accident investigation, etc.).

On the other hand, Harms-Ringdahl (1990) showed that large economic gains and related productivity gains could be made by introducing effective safety programs. He argued that absenteeism rates, sales rates and creativity levels are linked directly to the perceived level of health and safety care provided by the company. If the organization is perceived as interested and committed to health and safety, accident rates drop, and the before mentioned factors benefit. Safety, health, productivity and quality seem to be linked directly to each other as studies by Zink and Ritter (1993) show.

Furthermore, in the coal-mining industry Goodman and Garber (1988) demonstrated that absenteeism could not only be caused by higher accident rates, but also lead to more accidents, as constantly shifting teams of workers lead to more difficult communications, assumed agreements about working procedures which are not valid the next day and less care for team members.

It seems obvious that industrial accidents and occupational health are or should be a major concern for organizations and society as a whole and that human factors play a major role in the causation and prevention of such accidents (Hoyos and Zimolong, 1988). However, the specific role that risk taking behavior plays is not quite as clear.

7.3.1 Human Error and Accidents

Many statistical analyses of accidents and accident causes determine the human error, at the individual or organizational level, at about 80% while technical components only account for about 20% of accidents (Hale & Glendon, 1987; Trimpop, 1994 in press). Wagenaar, Hudson and Reason (1990), argued that accidents are the end result of long chains of events with managerial errors at the top (e.g. wrongly designed or not safety-oriented ordered machinery, unsuitable work procedures, incompatible goals, etc.). Errors made on the top hierarchical levels may remain hidden or being compensated for a long time until they interact with the necessary situation or behavior to create an accident. According to Wagenaar et al. (1990) many unsafe acts are not due to slips or simple errors but often intentional and reasoned actions that end in unforeseen results. The authors argued that erroneous plans are much harder to change at the shop floor level and thus, safety work should start and be most intense in preventive planning at the top level.

Human errors have been at the focus of attention of many studies and safety programs (Rasmussen, 1982). The most comprehensive analysis of human failures, with estimated probabilities of error and risk under specific technical design conditions and under stress has been performed by Swain and Guttmann (1983).

Often the methods and techniques utilized are expressed with letters (e.g., MAUT, HEP, SLIM, etc.) as synonyms. I shall present some of the major approaches and names as a point of reference for the interested reader.

Probabilistic risk assessments (PRA) are a systematic quantitative assessment of the likelihood of the levels of damage from operating industrial settings. Assessments are made by combining likelihoods of technical and human errors (Zimolong & Trimpop, 1994 in press).

One well-documented application of a procedure to obtain human error probability estimates in industrial settings is THERP (Swain & Guttmann, 1983; Miller & Swain, 1987). Other methods, such as the human cognitive reliability model (HCR, Hannaman et al., 1983) and the maintenance personnel performance simulation model (MAPPS, Siegel et al., 1984) have been discussed by Svenson (1989). A more detailed methodology has been used within the multi-attribute-utility theory (MAUT; Winterfeldt & Edwards, 1984).

The success likelihood index methodology (SLIM) was introduced by Embrey (1987). Human Error Assessment and Reduction Technique (HEART) was developed by Williams (1988; 1992) to account for interactions between performance shaping factors (PSF) while simultaneously incorporating a sizeable human error database and weighting factors that are based on extensive accounts of the human factors literature. The human error probability (HEP), used in many of those methods, is defined as the ratio of the numbers of errors to the number of opportunities for each type of error of interest in a task (Zimolong, 1990). The overall HEP of a task or sequence of tasks is calculated from the error probabilities of the individual elements of a complete task or the individual subtask. Basically, the individual task is broken into elementary units. For each unit the appropriate HEP is obtained from tables in Swain and Guttman's handbook, a decision is made if and how to modify the nominal HEP according to the influence of the most relevant PSFs, and the HEPs of the units are recombined in fault- and event-trees to model relevant parts of the complete system (Zimolong & Trimpop, 1994 in press).

In modelling human performance for probabilistic risk assessments, it is important to consider the factors with the strongest influence on performance. According to Swain and Guttmann (1983) some PSFs are external to the person, some are internal. External PSFs include the entire work environment, particularly the task and equipment design. Internal PSFs represent the individual characteristics of the person, his/her skills, motivations and expectations which influence performance. Other important PSFs are psychological and physiological stressors that result from a work environment in which the demands placed on the human by the system do not conform to his/her capabilities and limitations, such as task speed, task load and fatigue. These factors are approximated by estimation factors derived at through expert judgement techniques (Miller & Swain, 1987)

Hazard and operability studies (HAZOP) are probably the most common methods used for hazard identification in process systems and are also useful for complex manufacturing systems. The purpose of HAZOP is to identify all possible deviations from the way the design is expected to operate and all risks and hazards associated with these deviations. Fault trees, event trees, failure models and effect analyses (FMEA) are other reliability methods to identify inherent hazards and their consequences. A comprehensive overview on safety and reliability analysis techniques is given in Hoyos and Zimolong (1988), or Zimolong and Trimpop (1994 in press).

Major problems with these approaches are due to the fact that they largely rely on expert judgements to calculate weighting factors and that psychological stressors, such as inadequate managerial plans and decisions can hardly be integrated into such formulas. They may be useful for highly controlled, "stable risks", although even in such technological areas (e.g. nuclear power plants, chemical plants) accidents have shown that the interaction of human errors and situational irregularities can quickly overpower preliminary probabilistic risk assessments (Perrow, 1984; Dörner, 1989).

7.3.2 Influencing Occupational Risk Taking Behavior

As it seems impossible to design technologies error proof or even failure-tolerant enough to eliminate all risks and hazards, other approaches are being taken. Rules and regulations, for example, have been and still are commonly used methods of supervisors, organizations, and governments. Rules and regulations, however, have not proven very successful, as Hale (1990) showed. Rules seem to be more successful at a very specific level, provided the target group itself designed the required behavioral routines. A close inspection of human history also reveals that the enforcement of, or the identification with laws and regulations of a more basic nature than accident prevention laws has not proven successful. In spite of the clarity and widely spread knowledge, strong enforcement by government officials, commitment of supervisors and in spite of existing behavioral alternatives of the target group, even the *"10 Commandments"* have not been successful in making people adhere to the messages given.

Thus, the individual with his/her motivation (Spielberger & Frank, 1992), knowledge, abilities (Janssens et al., 1989) and possibilities (Zimolong & Trimpop, 1992) is at the focus of psychological organizational intervention strategies. Ultimately, behavioral components of the desired behavioral alternatives also have to be trained and supported, especially if unsafe or undesired behaviors have become automatic (Cox and Cox, 1991, Howarth, 1987; Sulzer-Azaroff, 1987;).

As mentioned before, in most cases occupational accidents do not happen because workers engage in behavior to experience positive adrenaline thrills, or to encounter a flow-experience through their actions (Wagenaar et al., 1990). Instead, many accidents, are for example stumbling accidents (30-45% of accidents in the chemical industry) that happen during ordinary walking from place A to place B within the company while they are talking or thinking of something other than walking. Other accidents happen when routines are changed without everybody being informed or when people underestimate the consequences (Musahl et al, 1992).

There are, however, instances when workers engage in dangerous activities for showing off as the most daring or most competent ones, and also situations where they are proud to utilize their improvisation skills (Trimpop, 1993). Many acts of unsafe behavior are, however, a result of the leading safety values, namely the safety culture and climate (Cox & Cox, 1991) or the perceived support and model of supervisors (Goldberg, Dar-El & Rubin, 1991). Other priorities are more important, such as finishing the job on time, keeping the engines running, or avoiding time loss by using improvised tools, such as boxes instead of ladders. In contrast to traffic participation, working seems to have many less telic (intrinsically motivated) motives and opportunities for physical risk seeking, rather more paratelic ones (Apter, 1984).

Due to the strong focus on productivity work places offer many opportunities and sometimes necessities for risk tolerance. Fire-fighters, policemen, coal-miners, truck drivers like many workers do have a higher "natural" or designed probability to encounter physical risks and accidents than for example working as an accountant or as university lecturer. Although there may be a risk seeking personality factor involved in the selection of a profession (Segal, 1993 and other studies discussed in chapter three), both questionnaire and interview data do reveal that workers largely deny engaging in dangerous actions to experience thrills (Hoyos, 1980).

A different picture is reported for managerial decision making. Data reveal here, that a substantial number of managers, at the higher levels of the decision and responsibility ladder, do prefer situations where they can make decisions that have significant consequences for the organization as a whole and thus for the livelihood of all its employees (Musolino & Hershenson, 1977). That way they seem to look for situations where they can prove their foresight, insight and ability to predict the unpredictable future better than their competitors. This behavior clearly falls under the definition of risk seeking or risk loving behavior. In contrast to people engaging in physical occupational risks, however, generally managerial risk taking is rewarded by promotion, more salary or a change of jobs, quite often regardless of the success of the risky decisions made. Thus, social and physical risk taking seem to be evaluated differently in work environments. Also, safety (or risk) laws and regulations do generally only exist to prevent physical damage and comparitevely few judicial regulations, such as a worker's council have been installed as a safeguard for social or financial risk taking on an organizational level. One exception in the last decade may be the increased public and, as a consequence, governmental interest in the protection of the environment and of public health (Wenninger, 1991).

Whether the target behavior relates to improvements of health, safety or the environment the first step in motivating people seems to lie in producing the perception of a problem or a personal threat (Goldberg et al., 1991). If people perceive threats then they are significantly more willing to participate in safety programs, provided they do perceive some control over the outcome. As already discussed in chapter two, perceived control seems to be a major contributing factor in risk taking behavior.

Harrell (1990) found that of 244 employed people those with the greatest perceived autonomy and control over tasks and work-pace also felt to be at the lowest accident risk. Generally, white collar workers perceived less risks in their working environment than blue collar workers. Both perceptions, however, were not related to actually experienced injuries. On the other hand, Jones and Wuebker (1985) showed, that people high in internal locus of control attributions think that they should take personal responsibility for their safety and that they can take preventive steps. Externally attributing people attribute accidents more to other people, circumstances or luck. Workers with an external locus of control also reported to have had more accidents than workers with an internal locus of control. Once a personal involvement, responsibility or threat is perceived, people are more willing to engage in safety programs. There are several psychological methods used to increase motivation for safety and safety behavior. Traditional approaches follow the theoretical assumptions provided by learning theory. Wilde (1976), Hoyos (1980), and Burkardt (1981) differentiate the following four strategies:

1. *Increase the benefits of safe behavior (e.g. praise, incentives)*
2. *Decrease the costs of safe behavior (e.g. time loss, comfort)*
3. *Increase the costs of unsafe behavior (e.g. financial punishment)*
4. *Decrease the benefits of unsafe behavior (e.g. recognition, time gains)*

While these approaches have shown considerable success (see Wilde, 1993 for a review on incentive programs), the methods are still based on hierarchical planning, orders and controls. Thus, the motivation is usually not intrinsic, the attitude towards safe behavior does not necessarily change and the behavior may be more oriented towards not being caught in performing unsafe acts, than towards self-responsibility.

Modern approaches to work safety strengthen intrinsic components of safety motivation, e.g. pride through participation in designing, performing and controlling adequate safety behavior (Zink & Ritter, 1993). *Safety Circles, Total Quality Management (TQM), Lean Management*, and similar organizational processes focus on the creative and participative involvement of all workers. Moreover it has been realized that zero-risk approaches are unrealistic, impossible, even undesirable. Workers are rather trained in accurate risk perception (Musahl et al., 1992), and behavior alternatives, in other words risk competence (Trimpop, 1994 in press).

7.4 Risk Taking and Traffic Accidents

Risk control through risk competence allows for seeking out and engaging in behaviors that provide challenges, opportunities for success and yet some control over possible adverse consequences. This combination seems especially important for professional and leisure time sport activities, but maybe very detrimental for traffic safety.

Traffic accidents clearly dominate the accident statistics, both for death, and for injuries. About 25 million people were injured in the USA and Europe during the 1960's (Svenson, 1978) and in Britain alone, In 1989 5373 people were killed and 336219 injured (Department of Transport, 1990). While Elander et al. (1993) estimated the loss of person-years for the USA to be about 1.4 million, studies assessing the exact physical, emotional and social costs of accidents are rare (Lave, 1987). Furthermore, even the individual amount of energy, mental load, physiological stress, etc. have only been examined in few studies. Elander et al. (1993) reviewed and summarized many studies on the effects of person-related variables, such as age gender, socio-economic status, personality, driving skill, etc. and came to the following conclusions:

1) Driving skill is related to accidents through speed of detecting hazards and switching attention, not however, through sensorimotor skills, or simple reaction time measurements, regardless of age, gender, and annual mileage.
2) Driving style is related to accidents through the relationship between general deviant behavior and traffic violations, maybe moderated through antisocial motivations and norms, and hasty decision making.
3) Stressful life events, such as divorce, may also play a role.
4) Young people, especially young males are strongly overrepresented in the amount of accidents incurred relative to their exposure.

7.4.1 Young Males and Traffic Accidents

Empirical evidence for multiple risk taking activities amongst young drivers has been found by Jonah et al. (1981; 1982). Jonah (1986) reviewed the evidence pointing towards a disproportionately high contribution of young males to accidents and concluded that young drivers are statistically at a greater risk of having an accident, and take higher risks in driving. This finding was supported by observational studies of driving behavior, as well as by attitudinal self-report measures.

Pelz and Schumann (1971) correlated accidents, age of driver, and distance travelled, and reported that age seemed to be a bigger contributing factor to accidents than mere experience. Von Benda and Hoyos (1983) reported that experienced drivers judge traffic risks more in a holistic way than younger and more inexperienced drivers. Wasielewski (1984) found that drivers from below an estimated age of 21 drove at an average of 76.5 km/h, while drivers estimated at a range from above age 21 to age 70 drove at an average speed of 70 km/h. Furthermore, Evans and Wasielewski (1983) calculated the distance between cars from photographs and estimated the age of drivers by the same means. They reported significantly shorter headways for younger drivers. Evans et al. (1982) and Evans (1991) found that those young drivers who did not wear a seat-belt, also drove at closer headways.

Aufsattler and Biehl (1986) reported, after inspecting all entries of traffic demerit points in Germany, that young drivers were strongly over-represented in committing traffic violations. They also found that drivers with multiple entries tended to repeat the violations that had given them the demerit points, and that drivers with severe first violations tended to have multiple entries in the demerit point registry.

Jonah, Dawson and Bragg (1981; 1982) reported that young drivers who admitted to riding their motorcycles after drinking alcohol were more likely to incur a motorcycle accident. The authors pointed out, however, that it is not known whether the same drivers who reported drinking and driving also had accidents, or whether the accidents were caused by alcohol.

Sivak, Soler and Tränkle (1989; 1989b) and Sivak, Soler, Tränkle and Spagnhol (1989) examined driver risk taking behavior in a computer-simulated driving task across three countries and found that apart from cross-cultural differences, young male drivers took consistently higher risks. The authors suggested, however, that due to their better physical and psycho-motor abilities these young males may have actually kept the overall risk constant.

Mercer (1989) also found young male drivers to have significantly more traffic convictions and to show more risk taking behavior in traffic. Mercer (1989) reported, however, that when the data were controlled for the higher exposure rates of young male drivers these differences became much smaller. In some instances risk convictions for young drivers were actually less than for older, and/or female drivers. Jonah and Dawson (1982) found, that younger drivers perceived themselves to have a higher risk of being involved in an accident in the next 12 months, and of being less likely involved in an accident than their peers (Finn & Bragg, 1982; Matthews & Moran, 1986).

Zuckerman (1979) reported that sensation seeking and risk taking behavior peak in early adulthood and level off after age 25. Zuckerman and Neeb (1980), and Clement and Jonah (1984) found significant correlations between self-reported and measured speeding on the open road. Colbourn (1978) also reported a relationship between age, risk taking, and simulated driving behavior in a video task.

7.4.2 Risk Taking Behavior and Motorcycling

Motorcycling takes a special place in accidents, as most riders are relatively young (Hurt Jr., Oulett & Thom, 1981) and as the riders use their vehicles for transportation, as well as for recreation (Hurt Jr. et al., 1981). Namdaran and Elton (1988) reported for a Scottish sample of 304 motorcyclists an accident rate of 8.2 per 100 riders in one year. Hurt Jr. et al. (1981) reported that 74% of 900 accidents examined were multi-vehicle accidents, and in only 33% of those the motorcyclist was at fault. The risk of having a motorcycle accident is therefore largely dependent on the abilities of other drivers, and only to a minor degree under the control of the motorcycle rider.

Also, in case of an accident, the chance of dying as a result is an estimated 25 times higher on a motorcycle than in a car (Hurt Jr. et al., 1981). These authors also reported that 75% of fatal accidents happened below a speed of 40 miles an hour (66 km/h) and 85% below 50 miles per hour (83 km/h). Only in 4% of all accidents were the motorcycle-riders above the legal limit for drugs, and 90% of the riders were not at all under the influence of drugs.

Schulz and Kerwien (1991) and Schulz (1993) examined the motivation of motorcyclists to engage in risky activities and the consequences for traffic safety. They found that younger drivers experience higher levels of control over their own behavior as well as that of other traffic participants. They also feel that their peers expect them to engage in risky activities and prefer sporty, faster motorcycles. They score higher on thrill, speed, and competition motivation than older drivers who score higher on escapism, hedonism and flow-experience (Csikzentmihaly 1985; 1988). Thus, young drivers are at an especially high accident risk, even compared to other motorcycle drivers. (see also Rheinberg et al., 1986; Rheinberg, 1991).

To summarize, young people, especially males, seem to be over-proportionally involved in risk taking activities, accidents in sports, as well as accidents in traffic. Although the data are somewhat confounded by the finding that young males have higher exposure rates than the average population, the fact that they voluntarily seek out these high risk situations still remains noteworthy. Furthermore, their motivation to engage in traffic or high risk sports seems to be strongly driven by emotional factors (Wagner & Vierboom, 1991). Therefore, it seems to be important to follow Jonah's (1986) suggestion, to determine reasons why young males engage in the above named behaviors, and whether the same people display risk taking behavior across situations and activities.

7.4.3 Risk Taking Behavior, Alcohol and Traffic Accidents

Noordzij, Meester and Verschuur (1988) reported that in contrast to low alcohol rates for motorcycle riders, for car drivers the rate is at least 20% of driving under the influence of alcohol above 20mg/ml. The low rate for motorcyclists may also be explained by the fact that driving a motorcycle requires considerably more balance and skill than driving a car (Wiegner, 1982; Hackenberg, 1985), so that drivers abstain more willingly from drugs that reduces their balance and psycho-motor skills.

Streufert et al. (1992) showed that visuomotor-performance deteriorated while under high levels of alcohol, while it stayed at placebo level at early stages on intoxication. Furthermore, only the error rate showed significant increases, while other measures, e.g. risk taking, seemed not influenced.

The relationship between alcohol and risk taking has provided mixed results. Contrary to popular belief that alcohol has a disinhibitory effect that facilitates risk taking, Adams (1988) argued that no conclusive evidence for any influence of alcohol on risk taking exists. He further argued that reductions in the alcohol related accidents would most likely not influence the overall accident rate in a population since it will just cause a shift in the type of accident.

Furthermore, even large scale legal measures, such as the 1982 Arizona Drunk Driving Law, did not have the predicted accident reduction beyond initial effects. Fatality rates first dropped in comparison to a control city but then slowly returned to baseline. Apparently drivers had adapted to the new enforcement practices and compensated in their avoidance behavior (West & Hepworth, 1989).

McMillen and Wells-Parker (1987), as well as McMillen, Smith and Wells-Parker (1989) reported that high sensation seekers took greater risks in a driving simulator than low sensation seekers, regardless of alcohol consumption. They reported further, that the actual amount of alcohol intake did not have any effect, but that the believe in alcohol intake did. High sensation seekers who believed to have consumed alcohol took more risks in driving in a simulator, while low sensation seekers took less risks when they believed to have consumed alcohol. The amount of risk taking was highest under moderate amounts of alcohol (McMillan & Wells-Parker, 1987). These findings would explain studies which have found an effect of alcohol consumption on risk taking, such as Mongrain and Standing (1989) as a psychological effect rather than a physiological one.

Joly and Wilde (1987), as well as Wilde, Trimpop and Joly (1989) also found no significant effect of blood alcohol levels above the legal limit on risk taking behavior in computer tasks. These results indicate that alcohol consumption has an automatic effect on psycho-motor skills, but not on risk taking strategy. Risk taking and alcohol seem to be connected through more complex psychological relationships, such as personality patterns.

Stacy, Newcomb and Bentler (1991) examined the relationship between problem drinking and personality patterns. In a structural equation modelling approach they examined the hypotheses whether personality has a direct effect (e.g. susceptibility) on alcohol intake, or whether there is a mediating (e.g. alcohol use in certain situations) or a moderating relationship (e.g. disinhibition). Results indicated support for all three assumptions, suggesting that they all play a role under different circumstances or for different personality patterns. Two personality factors emerged as strong predictors for problem drinking, namely sensation seeking and cognitive motivation. The study showed that straight forward simple assumptions about problem drinking have to be expanded and adapted to specific circumstances and personality patterns.

7.4.4 Stress, Risk Taking and Traffic Accidents

To assess whether physiological arousal correlates with perceived riskiness of traffic situations, Gstalter (1985) let subjects drive an instrumented car through Munich, Germany, while they were monitored for stress and arousal. He reported low correlations between various physiological measures, probably due to large individual difference in such physiological measures as heart rate or GSR. Self-reports and behavioral assessments showed significant correlations with the occurrence of arousal-stress in relation to risky situations, not however with the type of information processing. Stressful, high load situations were responded to by compensatory action, such as reduced speed. Self-ratings of risk correlated positively with physiological arousal responses. Muroya and Ikeda (1982) reported that risk behavior in driving correlated with heart rate increases. Mental effort alone showed no significant change. Evans and Carrere (1991) showed that urinary catecholamines of urban bus drivers were significantly correlated with traffic stress (e.g. congestions) and a lack of perceived control over their job, even after controlling for personal health and experience.

7.5 Theories of Risk Taking Behavior in Traffic

To engage in any type of sport, occupational or traffic activity means to elevate one's risk of encountering an injury, or even a fatal accident (Brown & Copeman, 1975). Using a vehicle in a goal-oriented (telic) way may mean that the driver just tolerates, or even endures the perceived risk. Activity oriented (paratelic) driving, on the other hand, clearly receives a lot of its rewarding character through the perception of risk (Rheinberg et al., 1986; Battmann, 1984). Drivers will therefore most likely seek out situations, or perform behaviors that increase the experienced risk (Battmann, 1984). This behavior is described in the motorcyclists' slogan:
"I don't know where I'm going, but I'll be there real fast!"

Accidents, however, seem to happen to motorcyclists, just as to car drivers, mainly when driving in the telic mode, such as to work, or to some other specific destination, rather than when driving for recreational purposes (Hurt Jr. et al., 1981). Also, shifts in accident rates have been observed. Svenson (1978) reported a significant reduction in fatalities per 100,000 km travelled for Western Europe and the USA for the period of 1970-1980. At the same time, the number of cars and kilometers travelled per inhabitant rose significantly. Wilde (1986) argued that after having accounted for the statistical differences and the oil crisis effect, no net-reduction in accident rates per time-unit of exposure can be found.

To understand why accidents happen, to whom they happen and what psychological processes can predict and ultimately prevent them, it is necessary to closely examine type, amount, and conditions of risk taking behavior in driving. Apart from risk taking are factors such as lack of skill and errors in human information processing responsible for accidents (Rasmussen, 1982; 1985). Accidents can be facilitated by situational settings, or personal factors, or an interaction (Svenson, 1978). Svenson (1978) argued that accidents happen, because drivers overestimate their skills, do not perceive the risk within a situation or planned action, or take risks deliberately. Similar assumptions may hold true for pedestrians (Zimolong & Gstalter, 1984).

McKenna (1982) split accidents as caused by either personality factors, or environmental-situational factors, or a combination of both. I shall review prominent psychological theories of risk taking in traffic, beginning with the predominantly personality-focussed approach of accident proneness.

7.5.1 Accident Proneness

Accident proneness is based on the observation that some individuals have more accidents than they would be likely to have based on statistical accident data (McKenna, 1982; 1983). The notion of holding certain individuals responsible for accidents goes back to a study by Greenwood and Woods (1919, in Greenwood, 1950), who studied female workers in an ammunition factory. They found that a small proportion of workers accounted for most accidents. The authors concluded that this was due to an inherent predisposition towards accidents of these specific workers.

More support for "accident proneness" was produced by Rawson (1944), who found that truck drivers with the most traffic accidents also had the most personal accidents. Tillman and Hobbs (1949) also found results in favour of accident proneness for taxi drivers. Personality structures for being accident prone were also found in Dunbar's (1966) "rebel" type person, being impulsive, low in self-control, adventure seeking, does not make deliberate plans, but acts spontaneously. Also, Dunbar (1966) argued that a low level and vulnerability of self-esteem are responsible for accident proneness.

To explain the process causing accidents and errors, Dunbar (1966) restated Freud's (1904) postulate, that we do not make errors accidentally, but rather that we subconsciously intended to make that error to punish ourselves for self-attributed mistakes, or unacceptable thoughts or wishes.

Rawson (1944), and Alexander (1949) applied a psychoanalytic explanation to accident proneness and attributed a subconscious motivation to commit errors to a desire for self-punishment as a consequence of feeling guilty. Greenwood (1950) even suggested to develop a measure to assess accident proneness through tests to exclude those persons from certain occupations whose high scores would predict them to be "accident prone".

One of the most authoritative studies on accident proneness used an extensive psychological test battery and tested the 66 subjects (bus and streetcar drivers in Helsinki) repeatedly over a period of 10-27 years (Häkkinen, 1979). Häkkinen (1979) reported that the correlation between having an accident in the initial testing period and 10-27 years later was 0.56, thereby indicating a high accident consistency. High accident drivers attributed their accidents more externally than low accident drivers, and differed in other features of the 18 different tests performed in this study. Häkkinen (1979) was able to apply a discriminant function analysis to the results of these 18 tests and he could classify 86.4% of the drivers correctly.

Another longitudinal study of accident proneness was performed by Dvorak and Linke (1976) who reported the relatively high correlation coefficient of 0.67 between two test points for accidents by 239 German locomotive engineers over a period of 16 years.

Further support for accident proneness was given by Platenius and Wilde (1989) in a study on accident involvement of 8800 pilots, who also showed either none or multiple accidents. Yet, the authors attributed their findings more to statistical artefacts or the work organization, than to individual shortcomings of the pilots.

Evans and Wasielewski (1982) observed 2576 drivers on a highway and reported that accident-drivers have shorter following headways than no-accident-drivers. Fernandez-Seara (1978) also reported that certain personality types had more accidents.

Synodinos and Papacostas (1985) designed a driving behavior questionnaire and reported recklessness in driving to correlate with the Type A personality depicting a behavior pattern with high levels of speed and impatience, as well as hard driving and competitiveness (Jenkins, Rosenman & Friedman, 1967; Jenkins, 1971; Furnham, 1984). This indicates that complex personality patterns are influential in "accident proneness".

However, despite some evidence for accident proneness, quite a few studies have failed to detect accident prone individuals in relation to accidents. For example, Leaman and Finch (1987) found no relationship between either *impulsiveness* or *venturesomeness* with accidents of young motorcyclists. Mohr and Clemmer (1988) found no indication for accident proneness in workers in heavy industry, and even after removing the few workers with several accidents, no appreciable drop in the accident rate occurred. Other findings cast some more doubt on a concept of strict accident proneness (McKenna, 1983), while authors such as Wenninger (1991) or Elander et al, (1993) even argued that there is neither statistical nor convincing empirical evidence for the concept of accident proneness. One of many arguments against a concept of accident proneness is that in a Poisson distribution of accidents, it is of course likely to find people with many accidents, just as to find people with no accidents (McKenna, 1982; 1983). Haddon, Suchman and Klein (1964) proposed that every accident is an interaction of safe or unsafe actions dependent on an individual situation and not an individual person. McGuire (1976) argued that their is reason to believe that those people who have accidents in one period of time, may not be the same as those who have accidents in another.

McKenna (1983) argued for a closer examination of the underlying variables and situational factors causing accidents, to understand the concept of proneness, if such a concept exists at all. Hilakivi et al. (1989) reported correlations between impulsiveness, adventuresomeness and other subscales of Cattell's 16PF questionnaire with risky driving and accident involvement of young drivers.

Another study showing the influence of situational factors on the drivers personality was performed by Mayer and Treat (1977) who reported that personality factors, such as impulsiveness and social maladjustment only lead to accidents and errors when combined with situational factors, such as: stress at work, at home, or depression. Van der Colk (1988) attributed a large amount of the accident causation to processing errors.

A similar assumption was made in the accident causation models of Hale and Glendon (1987), Rasmussen (1985) or Trimpop (1988). In a study showing support for the influence of mental activities in accidents, Hoyos (1988) referred to the mental load associated with driving, which becomes even stronger with stress due to perceived riskiness. Gulian et al. (1989) showed that in addition to driving-stress variables, driving-external factors, such as family stress contributed to risky driving, while Heron (1957) stressed the increased likelihood of accidents due to boredom.

These results show the great variety of factors contributing to accidents, some of which may be due to personality patterns, such as sensation seeking or impulsivity. A number of studies show that there is a tendency for some people to have more accidents than others. To attribute these increased numbers to accident proneness, however, is not fully supported by the literature. In spite of some serious confounds the concept can be useful by stressing the possibility that some types of personality may need different types of training to influence their risk taking behavior.

7.5.2 Traffic Accident Models Focussing on Undesired Risk

Näätänen and Summala (1974; 1975; 1976) referred to the subjective risks experience and discussed the effect of several "inhibitory motives" in different situations. These authors, however, stressed, that risk is only seldom experienced by traffic participants, since dangerous situations are relatively rare, and driving follows an automatic scheme, or habit. In their "zero-risk" model they suggested a theoretical construct called the "subjective risk monitor", which inhibits the experienced risk through its inhibitory function and influences behavior to meet expectations through its regulatory functions. As drivers typically want better performance, or to reach their destination earlier, they tend to choose higher speed, if the road and vehicle conditions allow them to do that (Summala, 1988).

In an experiment to test their assumptions, Näätänen and Summala (1975) let subjects throw darts at a specially designed dart board with a field labelled "near accident", earning zero points, and another area called "accident" where penalty points were awarded. On the remaining dart-board, highest points were earned closest to these fields. Subjects progressed from the safe, low earning areas to the risky ones, and regressed somewhat after being punished. These results were compatible with the *Zero-Risk Model*, as well as with utility theories. The zero-risk model falls completely short, however, in situations were subjects enjoy driving fast, or in complex driving motives. Zero-risk would in principle require the car driver to take the train or stay at home.

Restrictive laws, such as speed limits, were suggested by Summala (1988), as effective means of reducing accidents, because they would limit the motivation to drive fast (according to Summala). The model fails, however, to account for positive experiences of driving, or risk taking. It also fails to account for an interaction of the risk monitor and personality variables, experiences or expectancies (Van der Molen & Bötticher, 1988). Finally, interactions between delayed feedback about consequences of actions in relation to accidents are neglected in the model.

Fuller (1984; 1988) discussed risk taking in traffic in the form of a "threat avoidance" model. The model is called threat avoidance since Fuller (1988) stated that drivers largely respond to the possibility of a threat, rather than to an actual problem, and that drivers spend most of their time in avoiding obstacles and potential accident hazards. Avoidance behavior is largely performed after perceiving the hazard rather than anticipating it, thereby making punishment as a consequence of error largely ineffective, since the punishment for the error simultaneously punishes the avoidance reaction. In contrast, rewarding anticipatory error-free driving would be much more promising from a learning theory point of view.

Fuller (1988) argued that much of a person's traffic behavior is rooted in the person's conditioning history, and that many actions are not performed as a result of a conscious, deliberate decision process, but are rather made automatically. Just as the models described before, however, the *Threat Avoidance Model* can not explain the desire to drive, to speed or to enjoy the activity of driving at higher levels of arousal.

Klebelsberg (1982; 1984; 1987) differentiated between objective safety, subjective safety, and the resulting safety in traffic. Objective safety refers to the physical forces and conditions in traffic (e.g., speed, friction) while subjective safety refers to the emotionally, and cognitively experienced degree of safety in every situation. Because the individual cannot attain perfect information, there is a discrepancy between the objective-physical and the subjective-experienced risk, which is called the resulting safety. As long as the subjectively experienced safety stays on the same or a lower level than the objective one, no risk of an accident occurs. If the subjective safety is overestimated in relation to the objective safety, however, an accident is likely to happen. The finding that young drivers have a disproportionately high accident rate is attributed to their overestimation of their subjective safety. The higher number of accidents for elderly drivers is attributed to age-related performance decrements, such as poor vision, hearing, reaction time (Klebelsberg, 1984).

Klebelsberg (1982; 1984) suggested that drivers practice anticipation abilities and danger recognition. He pointed out the danger of teaching danger-coping strategies, such as courses in car-handling, as they increase the subjective level of safety, thereby reducing the resulting safety.

Some problems with safety oriented technical solutions have been identified in relation to marked pedestrian crossing which produced significantly more accidents per user, than unmarked crossings and observable versus not-observable curves, both observations probably due to too high levels of subjective safety, with resulting risky behavior (Klebelsberg, Pilsz & Schibalski, 1979). These studies supported Klebelsberg's (1982; 1984) warnings about increased accident likelihood if the subjective safety becomes too high.

Mahalel and Szternfeld (1986) put forward arguments against making the driving task easier, because that would reduce attention levels and increase the overall accident risk. A study by Rumar, Berggrund, Jernberg and Ytterbom (1976) found that cars with spiked tires drove significantly faster on ice than cars without spikes. These results are not favourable for traditional safety approaches, they do, however, support Wilde's (1978; 1982; 1986) risk homeostasis theory which will be discussed next.

Chapter 8: How Do We Adapt to the Desire for and the Control of Risk Taking Behavior ?

Pfafferott and Huguenin (1991) summarized the results of an OECD report (1990) on *"Behavioral Adaptations to Changes in the Road Transport System"* in the following way. They reviewed the existing empirical evidence and found safety measures reducing adaptations in many but not all measures on cars, roads or driver training (e.g., larger stopping lanes on highways had no negative adaptation). Thus, there seems to be no more doubt about the fact that behavioral adaptation in response to safety measures does occur! However, the degree to which this adaptation, or compensation occurs largely depends on the detectability, the personal experience, and the promotion and advertising of safety measures.

Most theories on risk taking, adaptation or compensation include a feedback loop that provides the driver with safety relevant information. Driver behavior is influenced both through immediate information (e.g., speedometer, visual inspection of the traffic condition), and through general information provided by mass media, advertisements and official announcements. The most effective ways of structuring such feedback, news, information, and marketing of traffic safety as well as some considerations for evaluating the effectiveness are given in Wilde et al. (1974), Pfaff et al. (1982), Evans (1991), Becker, (1993), and Huguenin (1993).

Wilde (1986) discussed both the negative and positive effects of mass media. Negative effects to traffic safety arise especially in those conditions when safety measures are publicised in such a way that they lure the driver into a false sense of security (e.g. anti-lock braking system); whereas positive effects emerge, when the underlying motivations and values of safe versus unsafe behavior are changed (Wilde, 1993b).

8.1 Risk Compensation Behavior and Risk Homeostasis Theory

The concepts of compensation and homeostasis have been used in many areas. As early as 1859 in physiology by Bernard, and then later by Cannon (1929; 1932) who coined the term homeostasis in relation to bodily processes, such as blood pressure, temperature, and respiration rate, which all maintain an optimal level (Apter, 1982). For each specific situation and task appropriate individual values exist, but overall they remain constant. This process of fluctuation around a target level is called *homeostasis*. If one is infected the temperature rises, if one runs the respiration rate increases. The values fluctuate around the new level. When the situation changes, they return to the base-rate.

Other researchers have also described homeostatic processes. For example, Jung's (1954) compensation theory of personality and Lewin's (1951) "quasi-stationary equilibria" both referred to compensation behavior and homeostatic fluctuations of a variable within certain boundaries. Apter (1982) gave the example of driving on a highway. Although the car is sometimes more on the left or right of the road the average route driven is fluctuating around the middle of the lane; the process is homeostatic. Lazarus and Folkman (1984) stated that it is necessary to determine conditions disturbing homeostasis, create stress responses, or restore homeostasis in danger appraisal situations.

Risk compensation and risk homeostasis function under the same general principles of feedback-regulated adjustment behavior, in relation to the organism's abilities and the situational circumstances. Some examples may illustrate this point: If you are driving on a highway and it starts to rain, or snow, what do you do? I predict that you will slow down, to reduce the likelihood of having an accident. If the sun comes out again, I predict that you will speed up again, to a speed at which you will feel comfortable.

Similar arguments hold for different vehicles, where you will probably choose a higher speed in a car with functioning brakes than one without, or for a broad, well paved highway versus a small, rough road.

The behavior to counteract for an increase in perceived riskiness of driving is called risk compensation. The level of risk at which a person feels best is maintained homeostatically, both in relation to emotional and physiological experiences. Unless this compensation behavior has already become habitual, a plan for compensation in response to a given stimulus has to be designed. In the following, I shall describe the development of risk homeostasis theory, being yet the best developed theory of risk compensation.

Taylor (1964) described driving as a self-paced task, governed by emotional tension and anxiety levels, which represent the transformation of the physiological experiences of risk, which then in turn influence the driving task in a closed-loop feedback process. Taylor's (1964) subjects drove a car around a specified route through a city. The subjects' subjectively experienced risks were measured through fluctuations in the Galvanic Skin Response (GSR). The GSR correlated highly (r=.61) with the number of accidents that had happened at those spots before, according to police records. The subjects did not know those accident statistics, and over the full time of the trip their GSR levels remained remarkably stable. Taylor (1964; 1976) concluded that drivers compensated via their behavior in adjusting risk levels toward a subjectively desired level, which is then reflected in the GSR. In other words, subjects have most likely perceived at the accident spot some actual or possible danger, and since they drove more cautiously in response to that their overall GSR levels stayed stable. This finding suggested that subjects are able to perceive signals about risky situations without conscious knowledge (also called "precognitive loops", Michon, 1989), and that drivers compensate with their behavior (e.g., speeding, slowing down) for undesired levels of perceived risk.

Later on Heino et al. (1990) replicated Taylor`s study in part and found convincing support for risk homeostasis theory, but with slightly different behavioral patterns in their subjects than in Taylor`s.

Following Taylor's study, Cownie and Calderwood (1966) formulated the compensation principle and showed in *"black spot"* treatment that technical safety features are rendered ineffective, as drivers employ a risk compensation feedback loop, which made the drivers slow down at the "black spot", but compensate for the lost time by speeding afterwards.

Wilde (1976; 1978) followed up on this concept of risk compensation and formalized it by designing a model of driver behavior. This model is based upon compensating actions, which are triggered by a perceived discrepancy between the observed and the desired level of risk. Wilde (1982; 1986; 1988) extended the concept of risk compensation to the one of risk homeostasis, since the key concept of his theory does not focus on compensating for risk per se, but on optimising the level of risk taken in a homeostatic way. In his *Risk Homeostasis Theory* **(RHT)**, Wilde (1982) referred to a fluctuating level, which matches only on average a certain target level of risk. He clearly distinguished homeostasis from isostasis, the latter meaning an invariate, constant level of risk, for which homeostasis is often mistaken (Wilde, 1988; Trimpop & Wilde, 1993 in press).

Wilde stated (Wilde, 1982, p.10):

"a person while driving a vehicle, while involved in an industrial task or recreational sport, or while making decisions in any other behavioral domain that may have implications for health or safety, is acting in a way that may be understood as a homeostatically controlled self-regulated process. At any moment of time the instantaneously experienced level of risk is compared with the level of risk the individual is wishing to take, and

decisions to alter behavior will be made whenever these two levels are discrepant. Whether the ensuing behavior will have the desired result of re-establishing equilibrium between the target level and the experienced level of risk, depends upon the individual's perceptual, decisional, and executional skills. However, it will be argued that these skill factors as well as extraneous interventions that provide a greater opportunity, but not a greater desire for safety and health, do at most have a temporary effect upon the level of subjective and objective risk. The only factor that appears to determine the long-term level of subjective and objective risk is the target level of risk, which, in turn, is dependent upon the individual's evaluation of costs and benefits of various action alternatives."

Wilde`s risk homeostasis theory is depicted in Figure 1 (reprinted with permission from Wilde, 1982). A person's expected utility of action alternatives determines the target level of risk, while the perceptual skills determine the perceived level of risk. Based on a comparison and addition a desired adjustment is designed. The actual adjusting action is determined by the person's decisional skills and action skills, such as the ability to handle the vehicle properly. The result of the adjustment causes, and is reflected in a changed environment. Through a process of lagged feedback the individual takes in a new perception and the process restarts. The system maintains a homeostatic risk level over time, while continuously fluctuating to compensate for environmental changes.

Thus, according to Wilde (1985; 1988), externally introduced safety measures, such as seat-belts and improved brake systems, can at best only have an accident reducing effect until this result is perceived by the individual driver. After finding out about the increase in safety, the driver will then compensate for this increase by transforming it into performance benefits, such as shorter travelling time, or higher comfort through paying less attention, or some other means of risk taking behavior.

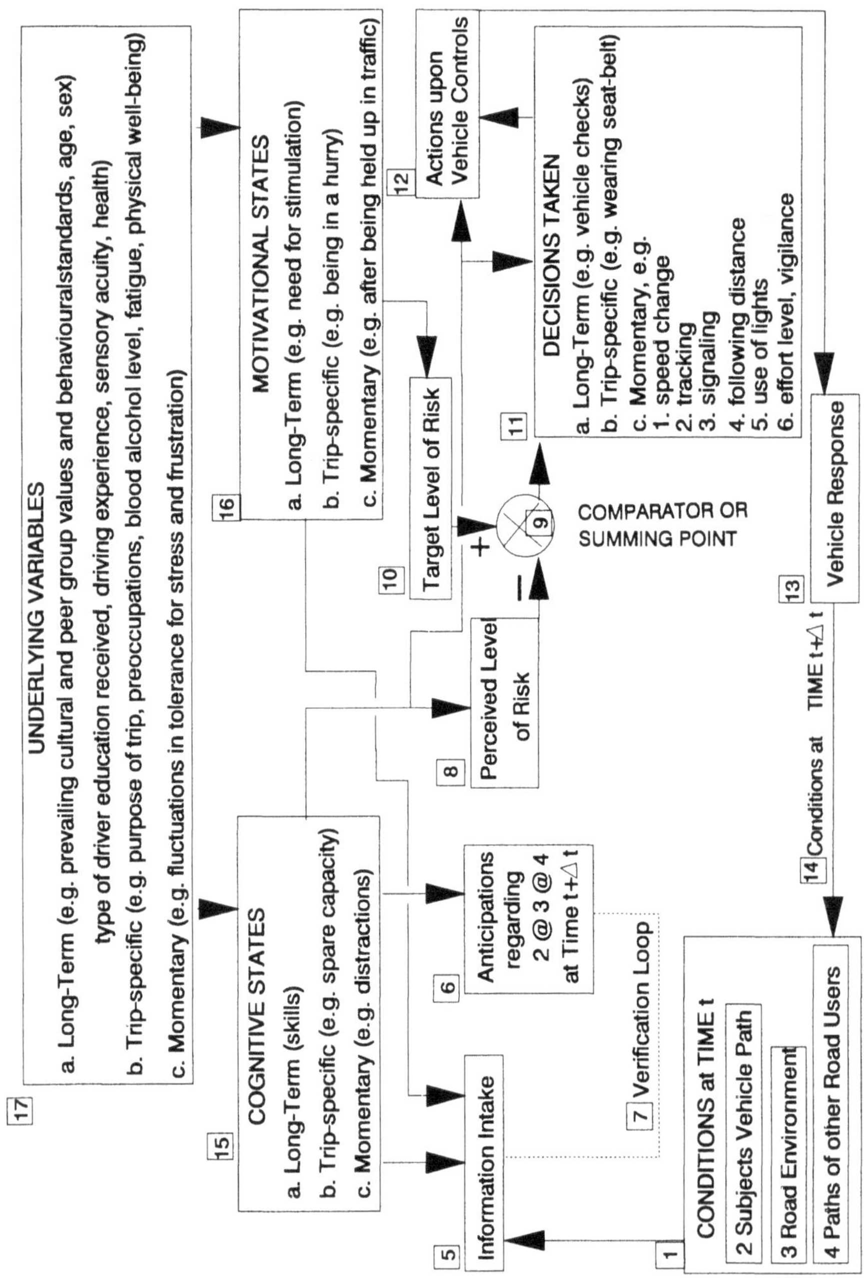

Figure 1: Risk Homeostasis Theory (RHT) in relation to driver behavior
(Figure is reprinted with permission from Wilde, 1982).

8.2 Technical or Psychological Accident Countermeasures?

A study by Peltzman (1975) was viewed by Wilde (1982) as supporting his postulate of the ineffectiveness of technical features. Peltzman (1975) reported no reduction in the accident rate per 1000 million vehicle miles, in spite of improved safety features, such as safety belts, impenetrable windshields, dual braking systems and padded instrument panels. Furthermore, traditional training and education programs have also received mixed reports about their effectiveness in reducing accidents.

Jonah and Lawson (1984) and Jonah and Dawson (1982) reported that the use of seat belts could be predicted from community and individual attitudes to the mandatory laws, and that the effect of these laws fell far short in their effect on Canadian fatality and injury rates. Svenson, Fischhoff and MacGregor (1985) reported that U.S.American and Swedish drivers overestimated both their driving abilities and driving safety, and that seat-belt use could not be influenced by advertisements about its effectiveness in reducing injuries, but it could be influenced by presenting facts related to personal comfort and social norms.

Lund and Williams (1985) reviewed many publications on safety courses and concluded that no support for their effectiveness can be found. Sometimes even a reverse effect has been observed, with the safety course drivers having more accidents afterwards. Kunkel (1983) discussed problems in studies of safety course effects, while Jonah, Dawson and Bragg (1982) referred to problems of self-selection, namely that mainly those motorcycle drivers participate in motorcycle safety courses, who are more safety conscious to begin with.

The significant drop in accidents after Sweden and Iceland switched from left-hand to right-hand traffic (Näätänen & Summala, 1976) can be explained by risk homeostasis theory. Due to the switch, a higher level of risk was perceived, thus the drivers compensated for this by driving extra carefully, until they got used to the new situation. This was marked by the return to normal accident levels over a period of one year in Sweden. In Iceland, with about 40 times less inhabitants it took only 2.5 months for the accident rate to return to its pre-switch level.

Wilde (1985) cited a study conducted in a German branch of the Kraft Food Corporation with a vehicle fleet of 600 trucks and vans. In 1957, drivers were told they would receive a bonus of about $150 for every half year of driving without "culpable accidents". This arrangement was made because insurance premiums rise if accidents are judged by insurance companies as the driver`s fault, while the insurance premium drops if the driver has no accident in a year. The program was still being continued in 1985, with remarkable reductions in accidents. In 1981 the culpable accident rate per kilometer driven was 14% of the 1957 rate (Wilde, 1985), while the total accident rate per km driven was at 25% of the 1957 rate.

Fox et al. (1987) examined the effect of offering trading stamps for no lost-time injuries, no equipment damage due to accidents, for making adopted safety suggestions, and for preventing an accident. Trading stamps were withdrawn when co-workers in the group got injured, or caused accidents with equipment damage. Long-term reductions in lost-time injuries and accident related costs were found in the two examined open-pit mines. Further reduction of accidents and accident related costs through the use of incentive programs, can be seen in a study by Siero et al. (1989).

Harano and Hubert (1974) formed three matched groups of 9971 subjects who received the promise of a one year extension of their driver's license without having to take the written exam if they stayed accident free in the next year. A second group with matched subjects constituted the no-treatment control group. Those drivers whose license renewal was up in the forthcoming year showed a significant reduction in accidents (22%) as compared to the control group and the drivers whose license renewal was coming up later; their accident rate remained unchanged. Especially effective was the treatment for drivers below the age of 25, being the group with the highest proportion in accidents and the lowest income. In the following years the reduction in accidents was 33% before the experiment was stopped. More support of the general effectiveness of incentives in the reduction of accidents was reviewed by Wilde (1992).

Strong support for risk homeostasis theory came from a study by Aschenbrenner and Biehl (1988; English version: 1993 in press). Over a period of 36 months they observed part of a taxi fleet in München (Munich), Germany. Half of the observed vehicles were equipped with an anti-lock braking system (ABS) that avoids the locking of the wheels in case of a hard braking manoeuvre. This mechanism preserves the possibility to steer the vehicle and improves braking efficiency, especially on wet roads (Aschenbrenner and Biehl, 1988; 1993 in press). The other half of the taxi fleet consisted of identical cars without the ABS-system. Drivers used the vehicles interchangeably, controlled by a random assignment system. In addition, acceleration changes were recorded inside some cars, speed checks were made on several spots of the city, driving styles were observed by unobtrusive passenger-observers, accident data gathered, and personal interviews conducted.

Aschenbrenner and Biehl (1988; 1993 in press) found taxis with and without ABS to drive equally fast while ABS taxis braked quicker. Drivers seemed to overestimate the abilities of the system. They believed that ABS would give their cars better braking power under all road conditions, and better car steering abilities, even if not braking. Most drivers did not know the technical features of the system. Many drivers (62%) stated that the

ABS-system makes people take more risks and drive less cautiously. In relation to accidents, Aschenbrenner and Biehl (1988; 1993 in press) found that accidents due to another car cutting in before the ABS taxis were significantly reduced, while other accidents, especially in snowy conditions, rose. The overall accident rate showed a slight increase for ABS taxis, but no significant differences between cars with the superior brake-system (ABS) versus the cars without the system. The authors reasoned that the results can best be understood by the theory of risk homeostasis, since compensation behavior and related psychological phenomena were assumed to have taken place in these areas:

a) Drivers drove less carefully, since they felt more secure;
b) Drivers overestimated and generalized the effectiveness of the brake system in relation to every situation;
c) Drivers received less feedback about the road conditions, as the braking system compensated for wetness, thus they estimated the road to be in better condition than it really was;
d) The drivers' belief in ABS brakes and their own superiority may have lea to a decrease in perceived personal responsibility, an increase in the target level of risk and a resulting increase in risky driving.

Data collected by Daimler Benz A.G. (1986) showed that vehicles with ABS were involved in accidents where they hit other vehicles from behind three times more, apparently due to smaller headways left by ABS drivers. Unpublished data by the German car company AUDI, show that vehicles with four-wheel drive, designed to improve the traction on the road in curves, were three times more involved in accidents due to excessive speed and seven times more in accidents gliding out of curves due to excessive speeds than identical vehicles without four-wheel drive.

These reports strengthened Wilde's (1982; 1986) postulate that drivers transform safety benefits into performance benefits, and show compensating

behavior, if they perceive lower risks. Technical safety improvements alone are thus ineffective for reducing accidents per head of population.
Further support for Risk Homeostasis Theory from a controlled experimental study was reported by Streff and Geller (1988). They let subjects ride go-carts on a racing course and split them into four conditions:

a) during the first trip only,
b) seat-belt on during the second trip only,
c) seat-belt on constantly, or
d) seat-belt not at all on.

The authors found a significant increase in speed for the group that drove first without a seat-belt and then put it on. The effect only occurred within subjects and only if they could directly compae conditions.

Jannssen (1988) conducted a test of risk compensation and incentives in a laboratory study. The following results were observed:

1) After a baseline of trials an incentive was offered conditional on accident free performance, resulting in a reduction of accidents by 11%.
2) The probability of an accident was reduced by 50% without telling the subject, which resulted in a gradual increase in the accepted risk levels until the 50% increase in safety was reduced to 20% through compensating behavior, when the experiment was stopped.
3) Under similar conditions but with the subject being informed about the 50% reduction, an immediate, total compensation was observed.
4) In a control condition no trends in risk-strategy shifts occurred.

Tränkle and Gelau (1992) compared RHT and SEU models in computer games, one of which estimated the risk taking strategy of subjects in manoeuvring cars across an intersection with gaps of different size. Risk homeostasis theory was a significantly better predictor than SEU, while in other experimental conditions the results indicated the opposite tendency.

8.3 Risk Homeostasis Theory Applied to Accident Reduction

Wilde's (1982; 1985; 1986; 1988) solution to the seemingly impossible task of accident reduction, lies in concentrating on the target level of risk, instead of on mere technical solutions. As long as the utility for risk taking remains unchanged, the motivation for compensation behavior will also stay at the same level. The target level of risk depends on the:

a) perceived benefits of risky behavior,
b) perceived costs of risky behavior,
c) perceived benefits of cautious behavior, and
d) perceived costs of cautious behavior (Wilde, 1982).

These utilities are a function of economic, cultural, and person-related variables and their long-term and short-term as well as momentary fluctuations (Wilde, 1986). A change in these variables is, according to Wilde (1985), the solution to reduce accidents, because an increased value of perceived benefits of cautious behavior, accompanied by a decreased benefit of risky behavior will result in a drop of the target level of risk. This means that the resultant behavior is based on a much lower desired risk-level (Wilde, 1982; 1984), which in turn will form the new target level of risk for the homeostatic process.

8.3.1 Financial Incentives to Reduce Risk Taking Behavior

Wilde (1985; 1989; 1992) proposed to use incentives to increase the benefits of safe behavior, while potential loss of incentives would increase the costs of risky behavior. Wilde (1991; 1993 in press) further argued that the *ups and downs* of economy do have a strong influence on accident rates, as poor economic conditions lead to reduced accidents, both because people drive less and more carefully. The importance of economic aspects in relation to accident rates was also discussed by Adams (1985). Adams

(1985) reported that the drop in accidents in the times of economic recessions were due to the lowered perceived benefits of driving in general, and of risky driving in particular. Both Wilde (1984) and Adams (1985) used data reported by Partyka (1984) to support their assumptions.

Partyka (1984) reported the number of people unemployed, employed, and not available for the labour force as economic indicators and compared this index to the fatality rate in the USA from 1960-1982. Applying her model to these data revealed that she could predict accidents from economic variables with a coefficient of determination of 0.89 for the general population, and for young people (age 15-19) of 0.98.

Adams (1985) used Partyka's approach on British accident and employment statistics and reported determination coefficients of 0.88-0.95. Economic fluctuations, such as unemployment, financial gains, or bankruptcy, can thus produce lower target levels of risk, because they increase the relative costs of accidents, and the costs of risky "joy-driving". This in turn, reduces the target level of risk and with that the motivation for engaging in risk taking behavior.

Similar results for other modern societies have been reported by Wilde (1993 in press). Wilde found correlation coefficients of up to r=.9 between the socio-economic status of a nation and its accident rate. Thus, when the economy is in a recession, unemployment and inflation rates are high and accidents are more costly relative to the reduced income, then the number of accidents are greatly reduced. At the time when this book was being produced, with a western-world-wide recession going on, accident rates should be significantly lower than in preceding years.

Wilde (1982; 1986) extended his theory from the domain of driving to other health-related life-styles and suggested that risk homeostasis theory is universal, but that there may be individual differences in the setting of the target level of risk. Wilde (1984) and Adams (1985) related risk homeostasis theory to other ill-health related life-styles by using data reported by Alderson (1981) who calculated the standardized mortality ratios due to violent death in 31 different nations between 1900 and 1975.

Wilde (1986) concluded that the death rates, after being adjusted for socio-economic status and age, stayed remarkably stable except during war times. The only country showing a great reduction in violent deaths was the USA, whose rate was, however, much higher to begin with, so that the USA are slowly approaching the significantly lower violent death rates of the other 30 nations. This extreme example of "death homeostasis" lends some credibility to Wilde's (1982; 1988) assumption of a population target level of risk, since a society ultimately sets its own standards and tolerance value, by which its members are educated, rewarded, or punished.

8.3.2 Open Issues in Risk Homeostasis Theory

To deal with all criticisms ever raised against risk homeostasis theory would require a book by itself, which is more up to the founder to write and is currently being done. Some of the criticism was based on misunderstanding, misreading, misthinking or mischief. I want to refer only to criticisms I regard as valid concerns and point to Wilde (1982b; 1984; 1985; 1988) for detailed discussions.

Adams (1985; 1988), Michon (1989) and McKenna (1990) referred to the difficulty in proving and disproving risk homeostasis theory, since the compensation behavior for induced safety measures can fluctuate over time and across activities. For obvious reasons, controlled laboratory studies can only control for certain behaviors and for a limited amount of time. Because of these methodological assessment problems, risk homeostasis theory can only be disproved theoretically, while support in favour of partial and complete compensation can be established empirically.

Cole and Withey (1982) argued that there are considerable difficulties in transferring individual motives and behaviors to a societal level, with within-subject empirical support specially designed to do that. Orr (1982) pointed out the difficulty for the individual of knowing probabilities of risk at the moment of decision making, but Wilde (1982b) referred to the notion of delayed feedback, so that the individual driver can utilize aggregate knowledge and constantly receives information through media, observations and by word of mouth to update personal risk assumptions.

Van der Molen and Bötticher (1988) showed weaknesses in most traffic safety models and suggested using a hierarchical model of risk taking. In this model they followed suggestions by Rasmussen (1982) to split behavior into knowledge based, rule based, and skill based. These levels correspond to the concepts of script, schema, and deliberate thinking.

Cognitive psychology uses the terms schema or script for these types of actions, thereby providing a hierarchical frame for risk taking decisions and behavior to be cognitively triggered either automatically, following a well learned pattern, or being deliberately planned.

If one follows a "script" - being an automatic, reflex-like learned activity - no time for evaluation and assessment of possible dangers remains. Scripts are the most commonly used action patterns, used for example in: speaking, walking, shifting gears, eating, or assembly line work. Such actions are characterized by an automatic, and often unconscious behavior. Scripts are overlearned responses, which can be retrained, altered or strengthened only by many repetitions or interrupting the automatic flow of action.

A "schema", on the other hand refers to a learned group of actions, procedures or other well learned, but not automatic actions. It leaves more chance to detect and respond to hazards, but as it is a well learned procedure, people are hesitant to interrupt it, and it requires mental effort to do so. Schemata are most often found in procedures like cleaning dishes, changing a tire, cleaning a workplace, climbing up a ladder, etc. They are characterized by an action which is not automatic, but requires only minimal attention. Schemata are learned and can be acquired through practice and deliberate thinking.

Van der Molen and Bötticher (1988) named their three levels *strategical, tactical,* and *operational.* At the highest level, route planning takes place, and the choice of speed, as well as decisions about which mode of transportation to take. At the tactical level, concrete manoeuvres are planned, such as overtaking, or accelerating. These plans are then carried out at the operational level through steering movements, or accelerator movements (Van der Molen & Bötticher, 1988).

Factors, such as personality variables, judgemental abilities, perceptual and motivational variables are incorporated. The model was then theoretically tested in the form of probability assessments and formal utility calculations. Strong experimental data have not yet been reported, and the area of emotional involvement, as well as physiological aspects and arousal seeking were not incorporated in the largely cognitive model. Adams (1988) also pointed out that it is very unrealistic in our society to call risk taking for risk's sake "pathological". The taking of risks by role models on television, in books, and in real life is heavily rewarded in all domains of private and professional life.

Lund and Zador (1984) did not find any differences in speed selection, stopping at amber, gap acceptance and following distance after the introduction of seat-belts. They concluded, that risk homeostasis theory did not occur and that safety belts will save many lives in Newfoundland.

Wilde (1988) argued, however, that the failure to observe changes during certain times of day does not rule out the possibility of changes at other than the observed times of the day or at night, and that the four observed behaviors are not the only ways one can drive more risky (e.g., more recklessly, less attentive).

Wilde (1988) cited data comparing Newfoundland and Nova Scotia for reasons of a control group, and showed that the type of accidents has shifted from the now presumably seat-belt protected car-drivers, to the unprotected other road users. In Nova Scotia, compared for the same years, the accident rate for non-car users dropped, until that province also introduced mandatory seat-belts, after which the non-car user fatality rate rose. Similar observations were made in Great Britain (Adams, 1985), USA (Peltzman, 1975) and Australia (Conybeare, 1980). It is, however, a valid problem, that risk compensation behavior is very hard to substantiate, since it can fluctuate across activities and time.

Shannon (1986) and Evans (1986) reported that the reductions by a factor of 2.5 in fatality rates from 1947-1972 could only be explained by RHT if the driving speed had doubled in that time. Of course the individual driver, or pedestrian, did not double the speed, but if one considers, and sums the increased mobility through more cars per inhabitant (0.23-0.58 times), more licenses (0.34-0.56 times) and greatly extended highway networks, the average speed per distance travelled for road users may very well have risen by a factor of 2.5.

Jannssen and Tenkink (1988) argued that phenomena observed as supportive for risk homeostasis theory can also be explained in a model of maximizing utilities, while disregarding risk completely. They included expected benefits and calculated the likely action for certain circumstances. As, however, RHT also claims to be a utility theory, it could be further developed to include the positive psycho-physiological experiences of risk, arousal, challenge, curiosity, as well as the intrinsic rewards connected to them. Jannssen and Tenkink (1988), however, did not consider non-material factors either.

Slovic and Fischhoff (1982) criticized risk homeostasis theory for not dealing adequately with the concept of *"need for variety"*. Slovic and Fischhoff (1982) claimed that need for variety and homeostasis are counteractive, as the searching for challenges, new experience and risks can not be explained by striving for a target level of risk, except after a process of adaptation, with a resulting raise of the target level of risk. Wilde's (1982) postulate that the target level of risk is influenced by motivators of various kinds, however, can account for those activities.

Wilde (1982) also postulated a consistent adjustment of behavior over time towards an optimal level of risk. As there are, however, large differences between the target levels of risk within the same motorcyclists while racing, or while wanting to read a book, it seems to be more appropriate to talk about multiple target levels of risk, which should fluctuate in response to every situational and personality change.

Wilde (1988) stated that a need for taking risks per se would be an unnecessary postulate, but that people instead choose their risk taking behavior in view of a certain subjectively valued goal. I argue, however, that one ***also takes risks for the intrinsic pleasures connected to risk taking***.

Choosing risks even in view of a goal must include intrinsic, activity oriented and non-conscious utilities as well. While not-goal-oriented motives are not specifically addressed in RHT their importance is acknowledged by Wilde in several publications:

"One does not travel only to arrive" (Goethe, cited by Wilde, 1982); *"psychophysiological needs"* (Wilde, 1985b, p.253); and:
"the not always rational appearing other utilities" (Wilde, 1988b, p.72).

Whether this behavior is called taking risks for the sake of risk, or for the intrinsic, partly automatic process goals is a semantic rather than a conceptual question. More arguments to include activity oriented risk taking and an intrinsic reward system for taking risks were presented in chapter three.

In addition, arousal seeking before and arousal reduction after reaching the optimal level appear to be rewarded intrinsically. Finally, people engage in many behaviors to avoid boredom, seek controlled thrills and risks (roller-coaster), are extremely curious, exploratory (a few hours with children will convince everyone), and engage in risk taking activities for the feeling of being alive, or to feel an arousal-high (Piët, 1987). They also engage in motorcycling in spite of, or even because of knowing the dangers.

I thus feel justified in suggesting that many of these behaviors mean taking risks more for the sake of experiencing intrinsically rewarding emotional and physiological consequences, rather then fulfilling a rationally derived decision to reach a certain goal, while merely tolerating risks.

If one argues, however, that intrinsic motivation is a goal in itself, then the term *"risk"* is excluded from being used as a valid goal, since it has no unique intrinsic or extrinsic value left to strive for. A similar argument was put forward by Janssen and Tenkink (1988b) who argued that risk compensation can be explained by utility functions. They stated that in most situations only partial compensation for safety measures takes place. The shift of risk taking behavior to other activities, or to other times, however, was not considered by them. Up to this date no study has been published that managed to control for many or all possible shifts in risk taking behavior.

Janssen and Tenkink (1988b) further advocated to differentiate between trip related versus situation related choice of risk in driving, similar to the separation of telic versus paratelic motivated risk taking behavior. Thus, if one refers to risk, one has to split behavior into the constructs of intrinsically and extrinsically oriented and motivated, or telic versus paratelic, or goal versus activity oriented behavior.

Strong support for risk compensation processes and mixed support for risk homeostasis theory has been found in laboratory experiments Tränkle and Gelau (1992). Carr (1985), Dickenson (1988), Hay (1988), Hunt (1986), Joly and Wilde (1987), Kuriyuchuk (1989), Morrison (1986), Parkes (1987), Rossiter (1987), West (1985), Wilde, Claxton-Oldfield and Platenius (1985), Wilde, Trimpop and Joly (1988), and Burns and Wilde (submitted for publication) have performed studies of risk taking in a computer game with various situations, such as alcohol, caffeine, music, competition, and sample differences, such as: sex differences, participating in a risky hobby, accident history, score on questionnaires, different psychomotor and cognitive skills, psychopathy, criminality, and being in the armed forces.

The methods used were described in detail in Wilde (1988c), but are essentially as follows: A light square on a video monitor had to be shut off as closely after 1500 milliseconds as possible. Too quick a response was punished in a probabilistic manner, with different probabilities to test for homeostatic shifts. Responding later than 1500 ms earned points in a decreasing fashion until 3000 ms, after which no more points were earned.

In addition, some of the above named studies used a trivial-pursuit type task under similar probabilistic conditions to account for task-related problems, as well as for testing levels of confidence. The results can be summarized as follows: The shift in the probability of penalty (POP) from being punished in 20% or 80% of the cases produced the corresponding shift in mean-reaction time. Under a lower threat of punishment the subjects made riskier responses. The optimal mean response time for each individual subject, considering individual skill level and the POP, was calculated and the deviation from this optimal score (DFO) equalled the subjects' risk strategy.

The assumption of risk homeostasis theory was that the risk-strategy, measured through DFO, would remain unaffected by changing probabilities of penalty. This was largely confirmed. Neither alcohol, nor music, nor caffeine had long-term significant effects on risk taking strategies, while short term fluctuations according to the level of penalty were observed.

Questionnaires, and accident histories did not correlate significantly with experimental measures, but the correlation coefficients were in the expected direction. Large individual differences in approach, strategy and reaction to the tasks were reported. Overall, the studies lent strong support to the concept of risk homeostasis, while leaving areas related to emotional experience and in-depth within-subject assessments largely open.

I conclude therefore, that risk homeostasis theory needs to expand its borders to focus on individual behaviors, on emotional, physiological and non-conscious motivators for risk taking behavior, and on constantly readjusted target levels of risk.

Chapter 9: How Can We Explain Risk Taking Behavior Holistically ?

After reviewing many different models related to risk taking behavior I came to the conclusion that each author was at least partly correct in his or her assumptions. Furthermore, most authors had convincing and sound research supporting their models. The more I studied the literature of the years 1860-1993 the more I came to the impression that we are just "reinventing the wheel", or that everything has been thought and said before. I am sure that many scholars have felt the same way, and yet, it looks as though some researchers seem to write as if they or their colleagues were either completely right or wrong, or as if they were the first or only ones to propose certain ideas.

It is hard for me to imagine that most researchers who have observed both their own feelings and behavior, and that of others, should be completely wrong, as representatives of apparently contradicting positions tend to argue. For example, the discussions of whether nature or nurture determines everything seems to be as futile, as the discussion about whether emotions precede or antecede cognitions, or whether non-material rewards are influencing behavior more or less than material ones.

If I look at my life, and that of the thousands of people I have known and consider their thoughts, feelings and behaviors in the large diversity of situations, then any "either-or" approach appears to be very simplistic and unrepresentative of life's complexity. In my experience, and I assume in other people's experience, there is ample evidence for some consistency in people they know across situations, and just as much evidence for the influence that different situations can have on the same person.

Furthermore, many of the decisions in my life were strongly influenced by emotions and non-material factors, such as love, honesty, friendship, self-worth, or a smile, while in other situations, subjectively rational, and material aspects played a strong role, for example in the decision to acquire a Ph.D. in Canada, or to structure the book in the way I did.

I want to advocate therefore that life usually does not consist of "either-or" decisions and situations, but that most life events emerge out of a mixture of wishes, emotions, skill, rational thinking, and luck. Similarly, I argue that risk taking cannot be explained as arising from a simple, clear-cut decision. People will neither take the highest possible risk in every situation nor avoid risky situations at all costs. There are, however, some tendencies for some individuals to be generally more daring or more cautious. Furthermore, people like stuntmen, astronauts, or race-car drivers have chosen a profession that is inherently more dangerous than that of a bank clerk or a psychology student. Within the professions, however, we find a similar split between a daring and a cautious chess-player or motorcycle-racer. The level of risk taking, or arousal seeking is different for the two groups, but the fluctuation within each level is probably very similar.

Going beyond my own experiences, the study of theories on risk taking let a similar picture emerge. Again, an either completely right or completely wrong approach seems to be insufficient for the complex concept of risk taking. At the same time, it may be possible to combine findings different theories have produced to develop a more complete, holistic model of risk taking behavior.

When I wrote my Ph.D. thesis in 1990 which forms the foundation of this book neither Frank Yates`s book (1992) on risk taking behavior nor the review paper by Elander et al. (1993) had been published yet. To my pleasure, in updating the literature and concepts for this book, I found that both authors had required theoretical assumptions and a holistic model for risk taking and accident causation. Thus, in this chapter I shall summarize the previously reported findings and propose a model of risk taking that combines major elements of the predominant risk-related theories.

I shall combine: theory of evolution, strength of the nervous system models (Darwin, 1859/1968; Gray, 1972; Pavlov, 1927; Strelau, 1974), optimal level of arousal theories (Easterbrook, 1959; Hebb, 1955; Zuckerman, 1979; 1991), opponent process theory (Solomon, 1980), reversal theory (Apter, 1982), utility theories (Slovic, Fischhoff & Lichtenstein, 1977), accident proneness (Häkkinen, 1979), risk homeostasis theory (Wilde, 1982), hierarchical risk taking model (Van der Molen & Bötticher, 1988), dynamics of action model (Blankenship, 1982), prime theory (Buck, 1985), affective reactivity (Larsen & Diener, 1987), desire for personal control (Burger & Cooper, 1979), and the two-process model of perceived control (Rothbaum, Weisz & Snyder, 1982).

The definition of risk taking behavior, given in chapter 1, requires that the here proposed theory of risk motivation must deal with all variables identified as important in this definition, as well as incorporate key components from the above named theories. To do that, I shall shortly present the components utilized in other theories related to risk taking behavior. Then I shall introduce ***Risk Motivation Theory (RMT)***. A schematic presentation of RMT will be followed by a description of the theory's components. Then, research questions derived from the model will be presented, leading to methodological design considerations for an empirical test of the validity of the individual components. Empirical research examining risk motivation theory will be presented and discussed in some detail at the end of this chapter.

9.1 Developing Risk Motivation Theory

In the preceding chapters I have reviewed theories and research in relation to risk taking. I shall summarize the major conclusions before attempting to integrate them into a holistic model, namely risk motivation theory.

9.1.1 Key Elements in Risk Taking Theories

Based on evolutionary theory (Darwin, 1859/1968; Dawkins, 1976; 1986) we would have to expect that people are generally both risk-inclined, as well as control-inclined. Risk taking is a necessary behavior to explore, cope with, and control an environment, the changes of which impose a constant level of uncertainty and danger on the individual. Control of the environment is necessary to reduce risks that are beyond the desired level and could pose a danger to survival.

The model by Buss (1988; see chapter three) described how important intrinsic emotional, physiological and cognitive reward systems are for risk taking and for a physical and psycho-social well-being of an organism, or a society. It was argued that too high risk motivation and too low risk motivation would prevent well-being, while optimal levels of risk taking correspond with high levels of well being. Due to migration, domestication and socialization, the levels of risk taking in the population should be distributed in form of a bell-curve.

Based on biological concepts, different individuals would have differently developed nervous systems, with different reactivity to arousal and risk taking (Eysenck, 1967; Pavlov, 1927; Strelau, 1972). Regardless of the level of preferred riskiness, however, people should strive to keep their level of risk taking as individually optimal as possible (Wilde, 1982; 1986). The optimal level of risk taking can be differentiated between an objectively optimal and a subjectively optimal level of risk taking, both of which may differ considerably between individuals, even if they perform the same activities (Zuckerman, 1979; 1985). It would be desirable to devise a measure of risk taking that determines risk motivation in relation to each individual's optimal risk taking level. People who take more or less risks than optimal against a certain criterion could thus be classified as risk seeking or risk avoiding.

The correctness of the chosen level of risk, however, can only be experienced and assessed by its consequences over time, such as, whether the organism survives, or experiences pleasure, pain, or well-being. The experienced level of risk taking can be evaluated by measuring self-assessed risk experience, and the physical, emotional, and cognitive well-being.

Blanchard and Blanchard (1984) explained that individual cost/benefit analyses determine well-being. These appraisals do not always necessarily happen on a conscious physiological and emotional level, as also automatic, intrinsic structures in the limbic system can execute appraisal functions. On the cognitive level, cost/benefit analyses are performed by conscious, rational decision processes (Slovic, Fischhoff & Lichtenstein, 1977). Such utility assessments depend on factors such as, internal/external reward type, probability of success, perceived control over the outcome (Burger & Cooper, 1979), perceived value of the activity itself and its outcome, expectations and experiences related to performing risk taking behavior, and related to the consequences of it (Fischhoff et al., 1981).

Discrepancies between a desired situation (target level of risk) and a perceived situation (or perceived level of risk) are reduced by compensating behavior (Wilde, 1986), or compensating cognitive reevaluation (Festinger, 1957; Rothbaum et al. 1982).

The motivation to engage in any special form of risk taking behavior is determined by anticipated intrinsic and extrinsic rewards, as well as by personal experiences with these rewards in the past (Buck, 1985; Heckhausen, 1988). Emotions and arousal influence each other directly (Larsen & Diener, 1987) or through opponent processes, which automatically reduce excessive levels of arousal and emotions (Solomon, 1980; Grossberg & Gutowski, 1987).

Depending on the interaction of situation and personality, a person will be either performing in a goal-oriented, or an activity-oriented manner (Apter, 1982), thereby displaying both state and trait related risk taking behavior. Furthermore, a person will actively design the risk taking activities in his/her personal and professional life to provide the best opportunity for experiencing as much arousal, challenge, and variety, as is individually desirable. The exposure to risk will be chosen such that both the automatic, intrinsic rewards of emotional well-being, and the secondary rewards provided through society (fame and fortune) are optimally high.

Differently difficult and differently risky situations require different behavioral responses. The responses have to be adjusted to the requirements in a "hierarchical" form (Van der Molen & Bötticher, 1988). Easy, or well known risks can be dealt with by automatic, operational scripts. More difficult or complex risky problems require tactical behavioral patterns such as schemata, while situations with high levels of uncertainty and complexity, require strategic planning of one's actions. These processes lead to different behavior depending on how a stimulus is perceived and by whom. Whether the person is more of a risk seeker, or in a risk seeking mode leads to different attributes than if the person is a risk avoider, or in a risk avoiding mode. The attributes listed in Table 1 have been suggested by different authors to describe risk seeking and risk avoiding:

Table 1: Risk Seeking Versus Risk Avoiding Personality Attributes

Risk seeking mode	***Author***	***Risk avoiding mode***
voluntary risk taking	(Slovic)	involuntary risk taking
performance oriented	(Dweck)	learning oriented
paratelic	(Apter)	telic
intrinsic motivation	(Deci)	extrinsic motivation
primary active control	(Rothbaum)	secondary passive control
problems are challenges	(Dienstbier)	problems are threats
arousal seeking	(Berlyne)	arousal avoiding
sensation seeking	(Zuckerman)	sensation avoiding
strong nervous system	(Pavlov)	weak nervous system
high emotional reactivity	(Larsen)	low emotional reactivity
high desire for control	(Burger)	low desire for control
high need for achievement	(McClelland)	low need for achievement
hope for success	(Blankenship)	fear of failure
risk enjoying, seeking	(Buck)	risk tolerating, avoiding

The physical properties of the nervous system determine the basic risk taking tendency and the basic target level of risk. However, the mode a person is in momentarily, will determine the physiological, emotional, and cognitive experience of risk taking, as well as the resulting decisions and actions. The mode depends upon personality factors, such as arousability, emotional reactivity, nervous system properties. The mode also depends on situational factors, such as problem perception, anticipation, stress interpretation, utility, and probability of outcome. Furthermore, whether risk taking will be experienced as physiologically, emotionally, and cognitively pleasant, is a function of the individual and total utility assessment of each component. The motivators for risk taking are therefore:

1) ***The desire to experience the inherent, intrinsic rewards for exploring and mastering one's environment.***

2) ***The extrinsic psycho-social and economic rewards for risk taking as provided by society (fame and fortune).***

In other words, the personality structure determines the basic target level of risk a person prefers, while the situation variables determine the momentary fluctuations. The actual risk taking behavior is therefore a result of the interaction between personality and situation variables.

To summarize: the following elements were identified as theoretically and empirically important to explain risk taking behavior and constitute the basic components of the here proposed model:

a) individual differences in risk taking motivation;
b) evaluation of situationally determined costs and benefits;
c) physiological, emotional and cognitive aspects of risk taking;
d) a behavioral compensation process to balance desired/perceived risk;
e) a dynamic feedback loop process model.

9.1.2 The Components of Risk Motivation Theory

Risk motivation theory is a dynamic state-trait model, combining personality and situation variables to assess the total utility of physiological, emotional, and cognitive cost/benefit analyses. Depending upon whether the total utility assessment indicates optimal, too low-, or too high risk, a motivation arises proportionally to adjust risk and arousal levels through continuing, or compensating behavior to establish or maintain an optimally pleasant experience. Results of the behavior are monitored in a feedback loop. The new perception will start the new risk-motivating process (see Figure 2).

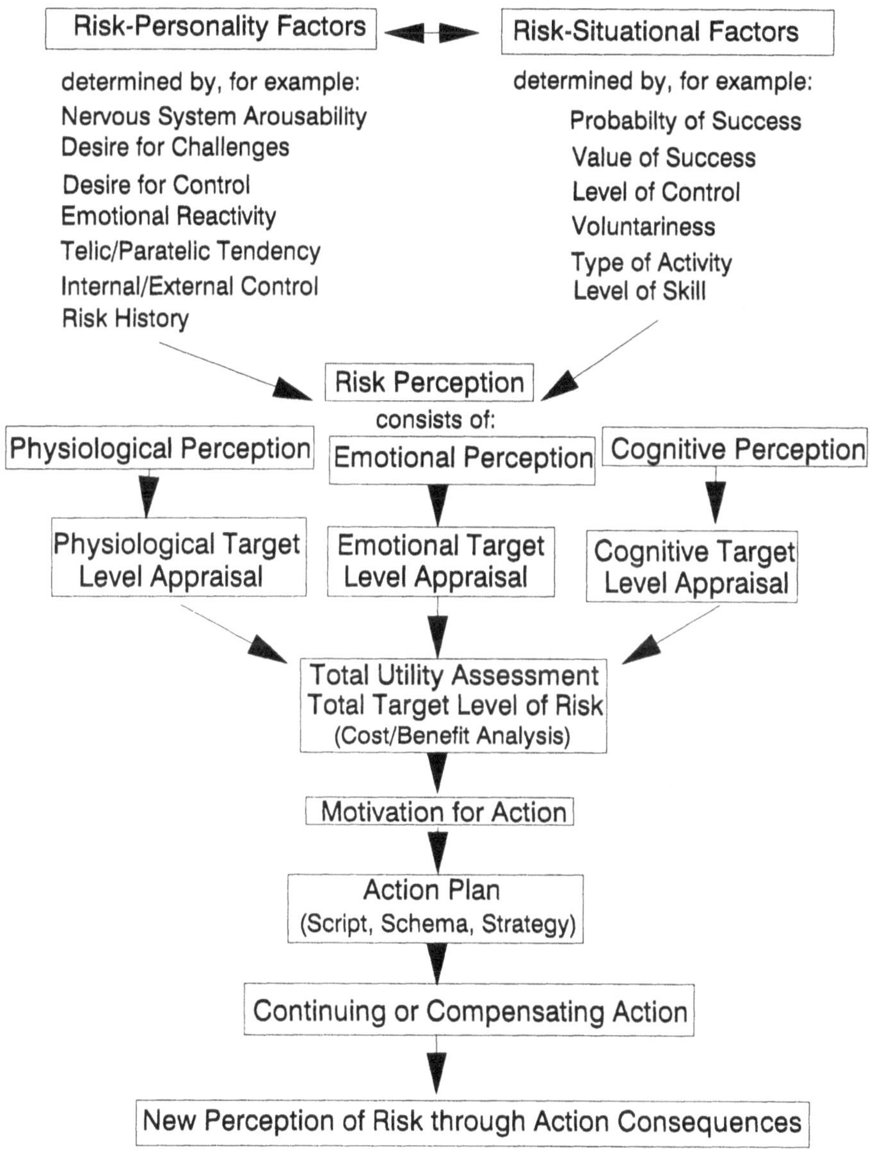

Figure 2: Risk Motivation Theory (RMT)
(Figure is reprinted with permission from Trimpop, 1990).

If the desired target level of risk and the perceived total utility level are identical, then the motivation to engage in risk taking behavior will be zero. Otherwise the motivation will require a new compensating action in either a risk-increasing or risk-reducing direction, thereby establishing a risk homeostatic feedback loop. In Figure 2, I presented the process schematically and next I shall explain the individual components of the model in their interrelations and consequences. Then, I shall link each individual component to assumptions that were empirically examined to test RMT.

Risk Personality Factors and Risk History

The first step in the process to determine behavior in a risky situation is the perception of a stimulus which sends an impulse through filter-systems to the central nervous system. Different risk personalities are influenced by physiological "responsivity", which sets the baseline for risk perception. Individually different physiological networks, sensitivities and synaptic connections are very likely the result of evolutionary selection. Situations with a constantly very high level of arousal, for example, should have favoured humans with a nervous system less sensitive to arousal by external stimuli. Lower arousability in the face of danger would lead to a higher target level of arousal, since it reduces the danger of overload and life-threatening panic. In return, different nervous system properties change risk-related personality traits, such as sensation seeking, desire for control, telic/paratelic dominance or emotional reactivity and temperament. *Risk Personality Factors*, thus influence *Risk Perception* in a hazardous situation.

A different risk history and connected associations can also cause different risk perceptions. For example, after having been attacked the night before in a dark street, seeing a person standing at a dark street corner may cause a drastically different risk perception, than if one had met a boyfriend or girlfriend at the corner the night before. However, individual differences in the basic arousability of persons should be preserved. For example, the 188 cm tall, and 90 kg heavy boxer should experience a different arousal level seeing the person in the dark street, than the 155 cm tall, 48 kg heavy school girl, regardless of previous experiences.

Part of this book describes the empirical support found for the risk personality components of RMT. Since risk personality factors are assumed to be relatively stable across situations and over time it can be postulated that individuals taking risks in one sphere of life should also take risks in other spheres of life. It is thus to be expected that recreational risk-activities, self-reported risk taking, vocational choices, accident and risk history, and experimental measures of risk show positive intercorrelations.

Situational Risk Factors

Temporary changes in the environment through *Situational Risk Factors*, such as skill, control, value and probability of success can also influence *Risk Perception*. Furthermore, physical factors, such as drugs, heat or other external factors can influence the perception and evaluation of risks. Changes in weather, mood, thoughts, hopes, fears, and the consequences of one's actions can also momentarily influence the perception of stimuli. Extreme heat or cold can make us respond differently to risky situations. The amount and quality of the processed stimuli determine any further processing action.

Since situational factors are normally of short duration, a withdrawal or ABAB design could prove the influence of such factors on risk taking behavior. It can thus be postulated that a change of the perceived value, perceived level of control, estimations of one's skill, and perceived riskiness of the situation, will have an influence on one's risk taking behavior.

Risk Perception

The perception of a stimulus is determined by the stimulus itself, the environmental conditions (*Situational Risk Factors*), and by the receiver (*Personality Risk Factors*) of the stimulus. The perception itself inervenes on different intra-individual levels: *Physiological Perception* relates to the physiological responses, such as hormones, blood pressure, heart rate, that automatically take place during and after the perception. *Emotional Perception* relates to the initial emotional perception of a process being pleasant or unpleasant, while *Cognitive Perception* relates to the first level of cognitive processing and evaluation of stimuli (*Apperception,* see chapter five). Cognitive perception also includes a categorization beyond the basic feelings of pleasant or unpleasant.

Physiological Target Level Appraisal

In the *Physiological Target Level Appraisal* the perceived physiological arousal is compared with the desired one. As a result of this comparison an immediate assessment whether the arousal is optimal, super-optimal, or supra-optimal in relation to a perceived stimulus is elicited. Extremely unfavorable, possibly damaging experiences, such as extreme heat, extreme stress, or extreme arousal will directly influence the total utility assessment, as well as the resulting behavior.

To assess what role arousal plays in relation to risk taking behavior and in the perception of risk, self-reported arousal as well as physiological correlates of arousal responses, such as heart rate and GSR have to be measured. It is assumed that arousal levels will fluctuate in response to perceived levels of riskiness. Furthermore, due to being more accustomed to risky situations risk seekers should respond with less arousal to the same experimentally induced risk than risk avoiders.

Emotional Target Level Appraisal

In the *Emotional Target Level Appraisal* all basic emotional experiences are compared to the desired feelings. The result is being processed further in a total utility assessment. The immediacy and strength of emotional perceptions assume a sensation, and - a not necessarily conscious - appraisal of the event. This can take place in the form of arousal levels being met, pathways in the limbic systems being used, or hormonal changes taking place. The output of this assessment can directly influence the total utility assessment and the resulting behavior. This appraisal ultimately serves the purpose to increase pleasurable and decrease unpleasant feelings.

The emotional experience connected to risk taking behavior can vary quite considerably, as shown in Apter's (1982) reversal theory. The same situation may elicit feelings of positive excitement or negative anxiety, depending on the personality tendency and the momentary mode an individual is in. Experiencing anxiety in reaction to risky stimuli, however, will most likely trigger different cognitive and behavioral patterns, schemata and scripts, than the patterns related to the feeling of being challenged. The perception of a risk is generally accompanied by physiological arousal and an emotional reaction of the immediate and automatic developmentally "older" brain system, as fight-flight reactions will be triggered quickly, and instinctively (Buss, 1986). Without feelings of uncertainty, or a certain level of arousal, a risk will hardly be perceived. Without an emotional experience of risk, however, only an abstract cognitive decision about a hypothetical problem can be made (e.g., the decision whether the probability of rolling a six on a die is as high as that of rolling a one). Laboratory research not eliciting emotional experiences of risk cannot be representative of real risk taking behavior, since the emotional variable is missing.

Furthermore, emotional experiences and physiological reactions, as well as behavior can be elicited via self-generated (without external stimulus) thoughts and fantasies. For example, the memory of an accident that happened some hours ago, may very well trigger emotions, arousal, and bodily reactions, such as anger, shivering, crying, or fright reactions. Emotional experiences in risky situations should be generally more pleasant for risk seekers than for risk avoiders. In addition, due to being more accustomed to risky situations risk seekers should generally show a lower level of emotions in response to experimentally induced risks than risk avoiders.

Cognitive Target Level Appraisal

Cognitive perception of a "risky" situation is influenced by physiological and emotional perception of this stimulus. If no arousal is noticed and no feeling experienced, a stimulus is not likely to be interpreted as "risky". Purely cognitive stimuli, like images or fantasies, should still elicit physiological and emotional responses, although weaker ones than those elicited by external stimuli.

The processing and conclusions of a *Cognitive Target Level Appraisal* depend on the experiences a person has had with similar stimuli, the perceived control over the situation, the perceived value or utility of the stimulus, the perceived voluntariness of the risk taking situation. The cognitive skill and decision making abilities are partly based on similar experiences in the past, or on previously learned coping skills. Furthermore, a low personal value of the perceived risk stimuli will likely elicit a "non-risky" assessment, and influence a cognitive assessment of the target level differently, than a highly valued outcome, or a highly aroused and little controllable situation would. This cognitive assessment, in return, influences the total utility assessment, the resulting behavior and the perceptions and appraisal of the outcome of the behavior via a feedback loop. It is therefore predicted that subjects will respond to changes in the perceived control, value of the rewards and probability of punishment by adjusting their behavior accordingly.

Total Utility Assessment, Total Target Level of Risk

In contrast to other utility models (e.g., prospect theory) of risk taking, risk motivation theory does not only take conscious cognitive cost/benefit analyses into account, but also physiological and emotional cost/benefit analyses. Physiological and emotional appraisals can receive more weight in the total utility assessment than cognitive appraisals. This represents the observation, that often emotional values or physiological desires can outweigh cognitive or rational ones. Drug users, smokers, gamblers, or the average person engage in behaviors because it *"feels good"* and in spite of better knowledge and although a rational decision would have suggested a different course of action. Furthermore, a strong threat of physical injury can elicit panic reactions, which in turn can prevent a person from following logical, cognitive problem solving strategies.

These aspects are evaluated when the three individual target levels (physiological, emotional, cognitive) are combined and compared to the desired *Total Target Level of Risk*. The total utility assessment of risk is influenced by differences in experience and personality. Cost/benefit analyses of the total utility assessment determine what combinations of individual target levels moderate the motivation for action. Manipulations of the probabilities of outcome, such as changing the objective probability of loss, associated with a given behavior, lead to changes in behavior, as compensating actions have to balance the desired and perceived levels of risk.

Motivation for Action, Action Plan

The aim for the *Motivation for Action* is to maximize the total benefit for the physiological and emotional evaluation, as well as for the rational assessment of the situation. Thus, as a result of the cost/benefit analysis, the total utility assessment either produces a desire to preserve the current state, or a desire to change it to a higher/lower level of arousal and risk. This motivation potential is the key motor for activating the corresponding action plan. This *Action Plan* may consist of overlearned scripts, or performance on the skill level, in Rasmussen's (1982) terms. It can also consist of schemata, or behavior on the rule based level, if the automatic scripts do not provide an immediate answer to the problem (Hale & Glendon, 1987). If neither available scripts nor schemata are sufficient to design an action plan, then a deliberate reasoning process on the strategy, or knowledge based level (Rasmussen, 1982) will be employed.

Depending on the perceived degree of difficulty of a situation and the motivation to change it, a hierarchically ordered action plan, similar to the one suggested by Van der Molen and Bötticher (1988), implements scripts, schemata, or strategies of compensating behavioral action. It is assumed that people who perceive large discrepancies between their target level of risk and their perceived level of risk will also have the highest motivation to adjust their behavior accordingly. The level of behavioral changes in response to changes in risk could be indicative for motivational changes.

Continuing or Compensatory Action, Feedback

The action that takes place in response to a perceived stimulus is either a *Continuing or Compensating Action*. This action is aimed at maximizing the total physiological, emotional and cognitive benefit by adjusting for shortcomings in either one of the three target levels of perceived risk. If the benefits of the situation are satisfactory for the physiological, emotional, and cognitive utility appraisals, then the behavior continues. The behavior may also aim at reducing the arousal perceived in the situation or at increasing it, if the individual is bored or under-aroused. The behavior is thus either maintaining or compensating for perceived physiological, emotional and cognitive costs/benefits to establish pleasant risk levels in a homeostatic fashion, similar to processes described in Wilde's (1982) risk homeostasis theory.

Performing an action changes the external environment of the acting person. The observation of these changes serves as information in a *Feedback Loop* to evaluate the success of the action plan. Results are perceived as a new stimulus, triggering a new compensation loop. The emotions, physiological experiences and cognitions of the first loop-process will now influence the physiological and emotional responses in the second loop-process. Cognitions, emotions, and physiological factors, thus have a continuous influence on each other, and can hardly be separated into precedences and antecedences. Similarly, personality traits, and situational states influence each other on all levels of risk taking and change the behavior constantly.

To summarize: RMT, a dynamic state-trait feedback model is aimed at explaining why people take risks. They take risks to maintain an optimal target level of physiological, emotional and cognitive well-being, as meeting these target levels is experienced as an intrinsically positive sensation. This intrinsic motivation is based on physiological reward structures and is of evolutionary origin. Optimal physiological, emotional and cognitive reactions and decisions maximize the chances for survival, as they provide warning signals, emergency support, and stimulation to actively change and master the environment for one's perceived well being. In addition, optimal performance is also rewarded extrinsically through financial and other material rewards for success.

9.2 Research Examining Risk Motivation Theory

As RMT describes risk taking motivation holistically, empirical research should also approach the examined subjects from as many different perspectives as possible. Thus, personality factors, situational factors, internal processes and overt behavior should be assessed both within subjects and across various risk taking situations and with changing conditions. The literature offers some methodological support in providing instruments that have been tested individually, but not yet in combination.

9.2.1 Methodological Requirements to Examine Risk Taking

For studying the relationship between personality traits and risk taking a number of questionnaires have been developed. In particular, sensation seeking, arousal experience, risk taking history, accident history, vocational choices, desire for control, overconfidence in skill and knowledge, and recreational risk taking activities were considerable aspects. Questionnaires should fit to the population to be studied and be relevant to the experiments performed. Since none of the reviewed questionnaires could provide ***all*** the essential information to answer the research questions related to RMT, a *risk taking history questionnaire* was especially designed. Research by March and Shapira (1992) also found risk takers` histories to be strong predictors of present risk preferences.

The influence of situational factors can be studied by changing reward characteristics of tasks, such as social rewards (Schurr, 1987), economic rewards (Ettenson & Coughlin, 1982; Janssen, 1988), or the threat of physical punishment (Paterson & Neufeld, 1987). Other aspects deal with changing the task itself (Wilde, 1988), the effect of time (Janssen, 1988), perceived control (Langer, 1975), or perceived skill (Wright, 1984).

A within-subject design with experimentally varied factors under controlled conditions should provide the opportunity for fair tests. The type of research best suited to studying situational influences, as well as changes due to reward structures and types of rewards, and the effect of situational changes on risk taking strategy and compensating behavior is found in studies on risk homeostasis theory. Examples of this research are found in studies by: Carr (1984), Dickenson (1988), Hay (1988), Hunt (1986), Joly and Wilde (1987), Kuriyuchuk (1989), Morrison (1986), Parkes (1987), Rossiter (1987), West (1985), Wilde, Claxton-Oldfield and Platenius (1985) and Wilde, Trimpop and Joly (1988). Most of these methods were summarized in Wilde (1988c) and Wilde (1993).

The physiological responses to risk taking situations should be assessed while the subjects perform their tasks, and should vary corresponding to the changes in risk. Physiological measures related to risk taking and arousal were also discussed in chapter 3, and the use of heart rate, or a measure of heart rate variability, combined with GSR was recommended, since both methods are easy to apply and cost effective, and have shown relatively reliable results in assessing the physiological correlates of risk taking (Grant, 1985; Näätänen & Summala, 1976; Van der Molen et al. 1987).

To assess emotional correlates of risk taking, Thayer's (1967; 1972; 1978; 1987) *Activation-Deactivation Adjective Check List (ADACL)* has been used to assess whether self-reported emotions about arousal and activity are related to risk taking behavior over time and in response to changing levels of uncertainty. Thayer (1978; 1987) reported that four factors of arousal responses varied across two dimensions, the factors being labelled: *General Activation, Deactivation-Sleep, High Activation, and General Deactivation.* Thayer (1967; 1970) found stronger correlations between the adjective checklist and combined measures of physiological arousal, than the intercorrelation between individual physiological measures. In addition, specifically designed questions should be utilized to assess those emotions related to the examined experimental risk taking behavior.

The influence of cognitive processes, such as perceived control, expectations about the outcome, cost/benefit analysis, and skill, have already been discussed under situational factors, but are best researched by assessing the risk taking strategy used in the Action Plan. The risk taking strategy can be determined by Wilde's (1988c) method of assessing each individual's deviation from an optimal risk taking behavior (DFO), under consideration of the individual's reward and skill characteristics.

Aspects of an individual tendency to gain mastery and control can be measured through Burger and Cooper's (1979) *Desire for Control Scale*, while the perceived *Locus of Control* in relation to driving can be assessed by the *Driving Internality/Externality Scale* of Montag and Comrey (1987). An influence of the perceived level of skill can be tested by measuring the correctness of one's degree of confidence in the correctness of answers in a cognitive task (Wright, 1984).

Continuing or Compensating Action (see Figure 2) to maintain a subjectively optimal, homeostatically controlled, level of risk taking is the result of a cost/benefit analysis of physiological, emotional, and cognitive motivations for personal well-being. The method to assess partial, total or even overcompensation is described in Wilde (1988c) and consists of changing probabilities of outcome, combined with perceived values, and their effect on risk taking behavior, and strategy. Since the strategy is postulated to be the achievement of an optimal level of physical and psychosocial well-being, the behavior should change, while the strategy should respond homeostatically. This process requires a feedback loop giving information about the outcome of one's behavior to make adjustments towards optimal benefits (Wilde, 1982; Lindeman, Van den Brink, & Hoogstraten, 1988).

One major problem in studying risk taking behavior is that of ethical considerations. It is not acceptable to expose subjects to grave risks for their physical, psycho-social, or economic well-being. Thus, although motorcycling ranks very highly in risk taking activities, it is hardly ethical to pay subjects for, or to encourage them to engage in motorcycle-races. Similarly, large economic risks, as incurred in stock investments, cannot be experimentally induced in the laboratory. Therefore, a compromise has to be found between the subjective experience of risk, which should be high, and the objective occurrence of risk, which should be low. Janssen (1988), Schurr (1987), and Wilde et al. (1985), suggested using economic incentives and social rewards, while Paterson and Neufeld (1987) suggested to use the threat of a physical danger, such as electric shock, to induce risk experiences without actually endangering the subjects.

Anderson and Brown (1984), Hudgens and Fatkin (1985), Whitaker (1980), and Wilde (1984) suggested to use game behavior and computer tasks as easily controllable risk taking behaviors. Arthur et al. (1990) even suggested the usefulness of computer simulated tasks to select accident prone truck drivers. Sorrentino et al. (1992) used Atkinson`s ring toss task and found subjects reported to experience feelings of uncertainty. In most cases the risks imposed consisted of physical, social, and economic rewards gained or lost with a certain probability. That way, any permanent harm is unlikely, while such rewards as money, threat of shock, or prestige have been successful in influencing risk taking behavior and danger appraisal (Paterson & Neufeld, 1987). To guarantee that the subjects experience risk at all, their self-reported risk levels should be monitored.

Furthermore, since risk taking takes place in many different domains of daily life, it should be tested across different domains of experimental conditions. Thus, different types of risk taking and different rewards for risk taking behavior, or risk avoiding behavior have to be used.

In the experiments described next, a within-between longitudinal ABAB design was used. The study consisted of the following four components:

1) A dice-rolling task where money was bet on the outcome,
2) a psychomotor, response-timing task, for financial and social rewards,
3) a cognitive, "trivial-pursuit" type task for financial and social rewards,
4) an electric shock question and six risk taking questionnaires, namely:

- Sensation Seeking Scale (Zuckerman, 1979);
- Telic/Paratelic Dominance Scale (Apter, 1982);
- Risk Taking Inventory (Keinan et al., 1984);
- Desire for Control Scale (Burger & Cooper, 1979);
- Driving Internality/Externality (Montag & Comrey, 1987);
- Risk Taking History and Risky Recreational Activities.

The participants were 120 males recruited from the City of Kingston, Ontario, Canada, with an age of 16-30, all holding a valid driver's license. Each subject participated in every component of the experiment, however in randomly assigned, different orders of presentation. In addition, the dice-rolling, cognitive and psychomotor tasks contained experimental manipulations regarding type, probability, and level of reward and riskiness. Changes during the performance of the cognitive and psychomotor task were being monitored in relation to physiological responses of the subjects' heart rate and galvanic skin response. After each change in the experiment, a questionnaire was administered to assess emotional reactions to the changes.

I shall describe the procedures, results and a short discussion seperately in relation to each of the four experiments before discussing the study as a whole to assess its validity for a holistic view of risk motivation and risk taking behavior.

9.2.2 Dice Rolling Experiment

After reading the instruction sheet subjects were asked to roll the die into the opening of a box, and bet between 0 and 25 cents on the roll being either "low" (1-10), or "high" (11-20) on a 20-sided die. Subjects were told that they could bet up to 25 cents in 5 cent steps for each of their 50 rolls. Whatever fraction of money out of the 25 cents they would not bet they could keep, whatever money they did bet could be either doubled or lost, depending on the correctness of their "high", "low" prediction. Subjects were given the money container and told that if they would not bet at all, they could keep $12.50, while if they bet all of it they could double the money, if their predictions were always correct. Then the subject rolled the die, took the money out of the 5-cent container and placed it in the betting container. Next, the subject checked the correctness of the prediction by lifting the box and looking at the die. The experimenter recorded both the amount of money bet and the amount won or lost. The same procedure was repeated 25 times.

This "betting after the roll" condition aimed at creating a subjective feeling of low familiarity, low perceived control, and high uncertainty, since the roll could not be influenced anymore, and thus more uncertainty was created in this condition. It was assumed that subjects would bet less money in this condition than in the condition of high familiarity, high illusionary control and low uncertainty. The condition of betting before the roll was also prevalent in 25 rolls. This condition resembled more the situation present in regular dice-games were the predictions or bets have to be placed before the roll, and the roll is not into a box. This condition should thus induce less feelings of risk and higher levels of illusionary control. The "bet before roll" group received identical instructions with the exception that the betting-before and betting-after order was reversed. Subjects were then lead into the adjacent room to continue with the study.

Results of the Dice Experiment

The assumption that if a situation creates the subjective perception of less personal control or more incurred risk, even if the perception is illusionary, people adjust their behavior by being more cautious and bet less money, was tested by comparing the average amount of money subjects bet.

In both situations the actual outcome was entirely determined by chance; subjects won on average \$12.42, with \$12.50 being the expected value. Subjects had been randomly assigned to either starting with the bet being made before the roll, or the bet being placed after the roll. An ANOVA was performed on the amounts of money bet. A significant difference between the two bet-order groups was found. Within the bet-order groups the condition of betting before or after the roll did not show any significant effect.

The results support the assumption, yet in a somewhat surprising way. No significant difference between the first and second 25 rolls was found. The prediction that subjects would always bet less money when the betting occurred after the roll was not supported either. Instead, this assumption was only found to be correct when subjects started off in the betting after the roll condition, as a significant bet-order effect emerged. The average amount of money bet suggests that people, when presented with the task of betting money after rolling, bet on average less money in this situation (368 cents on average for roll 1-25 in group A), as well as in the following situation of placing a bet before the roll (374 cents on average for roll 26-50 in group A). On the other hand, when subjects started off with the condition they were supposedly more familiar with and experienced less uncertainty, namely betting before the roll (422 cents on average for roll 1-25 in group B), the amount of money bet was significantly higher. This higher amount of money bet stayed at that level even in the following situation of betting after the roll (423 cents on average for roll 26-50 in group B, see Figure 3).

Thus, the findings did not show the effect reported by Burger and Cooper (1979) that subjects high in risk seeking and desire for control are more subject to the illusion of control and thus bet less money after having made the roll. They only do so when they start off with the higher level of uncertainty condition. Thus, the result of this study suggests that subjects adjust to the degree of uncertainty they are presented with, and that this behavior shows a carry-over effect even if the degree of subjective uncertainty is presumably changed. Personality variables were not related to the money betting behavior. Future studies should examine this effect under completely balanced designs with money and without.

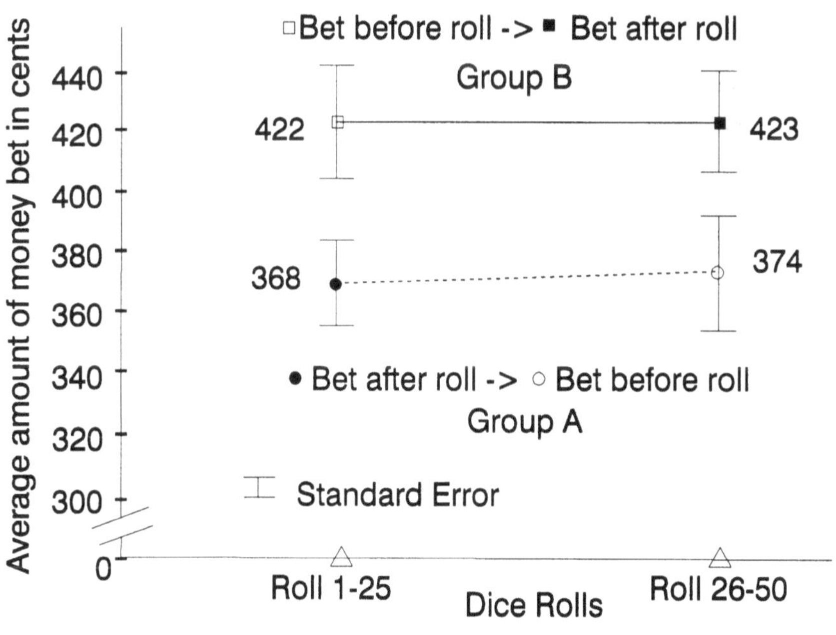

Figure 3: Group differences in the average amount of money bet
(Figure is reprinted with permission from Trimpop, 1990).

9.2.3 The Psychomotor Experiment

In the psychomotor task subjects had to cancel a stimulus light that appeared on a computer monitor as soon as possible after 1500ms, following its onset by hitting any key on the computer keyboard. Responses faster than 1500ms were punished in a probabilistic way with chances of either 20% or 80% to receive 0 or -5 points. Responses between 1500ms and 3000ms earned sure but linearly decreasing points. Subjects received information on the monitor before each block of 25 trials about the probability of penalty (POP) and feedback after each individual trial about response time, payoff, total score, and trial number. Mean response times were recorded as one dependent variable. The function depicted in Figure 4 was explained and always visible.

In ten practice trials subjects had a chance to become familiar with the task and ask further questions. Next, 12 blocks of 25 trials under either the social or financial reward condition were presented. The first four blocks of 25 trials were given at a probability of 80% of receiving -5 points, in case of a too fast response. The probability of penalty in these four blocks remained at 80% (=condition A). The next 4 blocks of 25 trials had varying probabilities of penalty, namely either 80% (=A) or 20% (=B). Thus, the blocks were presented in an ABAB, or withdrawal design. Each point was either added for the newspaper ranking list (social reward) or it was worth two cents (financial reward).

The final four blocks of 25 trials were identical to the ABAB design employed before, with the additional feature that subjects could double their scores, and thus double either money or ranking score, for each block in which they did not have a single penalized trial. It was emphasized that this *Double Reward (DR-ABAB)* could still be gained in case of a response quicker than 1500ms that received 0 points instead of -5 points. The distinction was made to see whether subjects would still compensate with their mean response times for the different probabilities of penalty of 80% and 20%. Following each block of trials the emotional questionnaire was presented. After completing the 12 blocks of 25 trials (300 trials in total) the cognitive experimental task followed.

While the psychomotor task was performed heart rate and GSR measures were measured and automatically recorded. A direct measure of heart rate variability was not possible with this method, but the sum of the absolute differences in heart rates after two consecutive trials (heart rate difference) was used to assess physiological responsivity.

After each block of 25 trials subjects had to answer the emotional questionnaire, consisting of the 20 adjectives of Thayer's (1967) ADACL, and seven personal experience questions asking for risk and arousal ratings in the last block of trials. Questionnaires were presented automatically on the screen and answered by hitting the corresponding numeric key. Subjects were constantly informed about their success in the tasks.

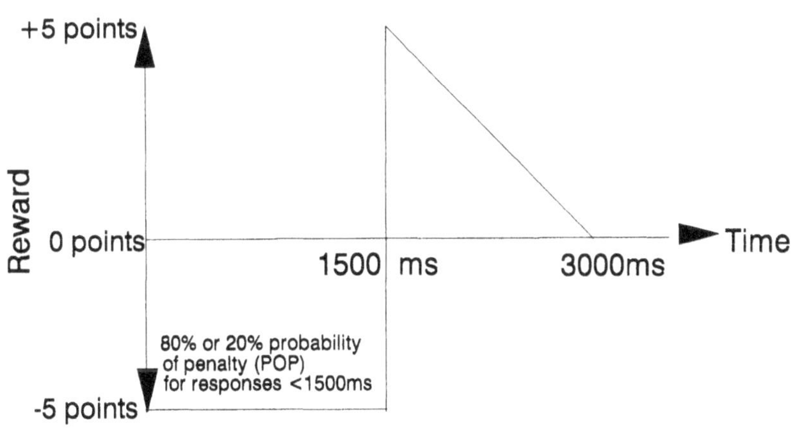

Figure 4: Response time - reward distribution
(Figure is reprinted with permission from Trimpop, 1990).

As the response times of each subject are determined by both their skill in time-estimating and their risk taking motivation, a measure of risk taking strategy was calculated that excluded the confounding skill factor. Skill can be expressed, in this experiment, as the average deviation from the subject's mean response time. Ideally, the mean response time should be always exactly 1500ms to obtain a maximum of 5 points at each trial. Since this is, however, impossible to achieve, the scores will be scattered around 1500ms. Furthermore, since too fast responses are many times more costly than slower responses (e.g., 1499 ms=-5 or 0 points; 1600ms = +4.67 points), the actual mean response time for each subject is expected to be somewhat above 1500ms.

This mean response time essentially depends upon two factors: First, the subjects' ability to estimate time accurately, and second, the subjects' individual target levels of risk. The closer the chosen response time is to the 1500ms boundary, the higher the likelihood for being penalized. The lower the skill of a subject to cancel the stimulus at the desired point in time, the higher the chance of being penalized. At the same time, however, it is to be expected that the increasing reward (more points) while approaching the 1500ms boundary will pose an incentive to the subject to gain maximum points, and thus, take risks. The mixture of skill and risk inclination complicates the differentiation between the risk taking tendencies of skilful, low risk-inclined subjects with a mean response time close to 1500ms, and, on the other hand of subjects with poor skill, but high risk-inclination, and thus an identical mean response time.

Therefore, a measure of risk taking has been designed that excludes skill differences between subjects (Wilde, 1988c). The measure designed is the subject's individual deviation from the optimal response pattern (*DFO = Deviation from Optimality*) in relation to maximizing the score. The DFO is derived at in the following way: First, each subject's *actual* distribution of response times with the resulting scores is determined. Next, the *optimal response pattern* for a subject is calculated. The optimal response pattern is based upon shifting the actual distribution of the responses (under consideration of the probability of penalty) on the response time scale until the payoff maximizes. It is thus determined how many milliseconds too fast or too slow the subject's actual response pattern deviated from the personally optimal response pattern, while leaving the actual distribution of responses intact. If the actual function was left from the optimal one (too many too fast responses) the subject had taken too high risks; if it was on the right side of the optimal response pattern, the subject had taken too little risk and responded too slowly for the personal skill and the respective probability of penalty.

This way, differences in individual skill become irrelevant and risk-inclination is expressed by the amount of milliseconds a subject deviated from an individual optimal response pattern. The DFO scale is a ratio scale of risk-inclination relative to risk neutrality. A deviation of 20ms versus 40ms translates to twice the risk-inclination. High risk taking is indicated by positive DFOs (too many too quick responses), while low risk-inclination is indicated by negative DFOs (see Figure 5).

According to risk homeostasis theory the DFOs are expected to come closer and closer to zero, if given enough time, feedback and practice, since subjects will tend to optimize their behavior in relation to their skill. In addition, according to risk motivation theory, any change or interruption of the task leads to a new assessment of the situation and the risks involved. Thus, a change in rewards lets subjects start again at higher DFOs before approaching the optimal deviation level of zero. Thus, if risk motivation theory is correct, DFOs should start again at almost identical points for the social and financial conditions.

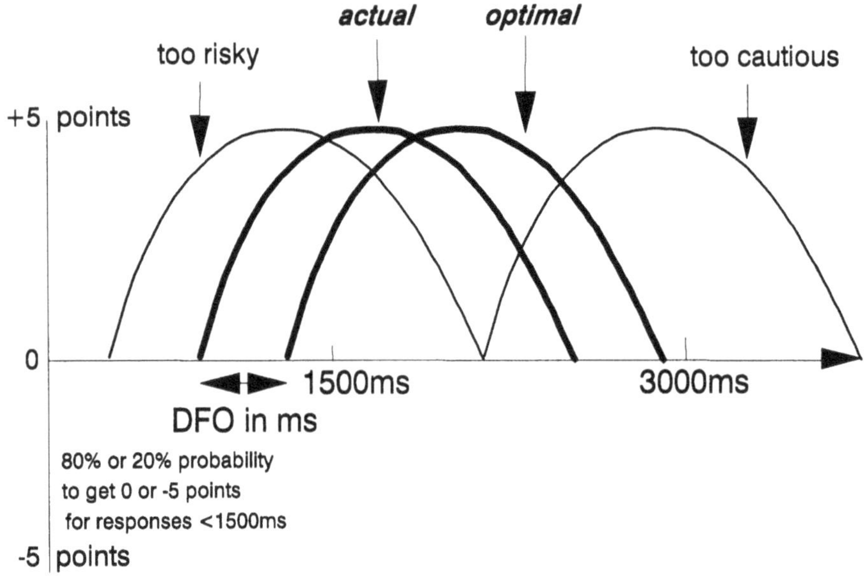

Figure 5: Risk inclination measured as the deviation in ms of the subject's actual
response time from his/her optimal response time (DFO).
(Figure is reprinted with permission from Trimpop, 1990).

Results of the Psychomotor Experiment

To test the influence of situational changes on risk taking trend-analyses of variance were performed on the measured variables, as in an ABAB design it is important to show that there are systematic fluctuations over time that correspond to the ABAB manipulations. The mean response times fluctuated strongly respective to changes of probabilities of penalty (POP). More risky situations were compensated for by choosing slower mean response times. Offering rewards for safe behavior (*Double Reward* ABAB-condition) slowed the mean response times even more down, while still maintaining the response to the ABAB experimental manipulation (see Figure 6).

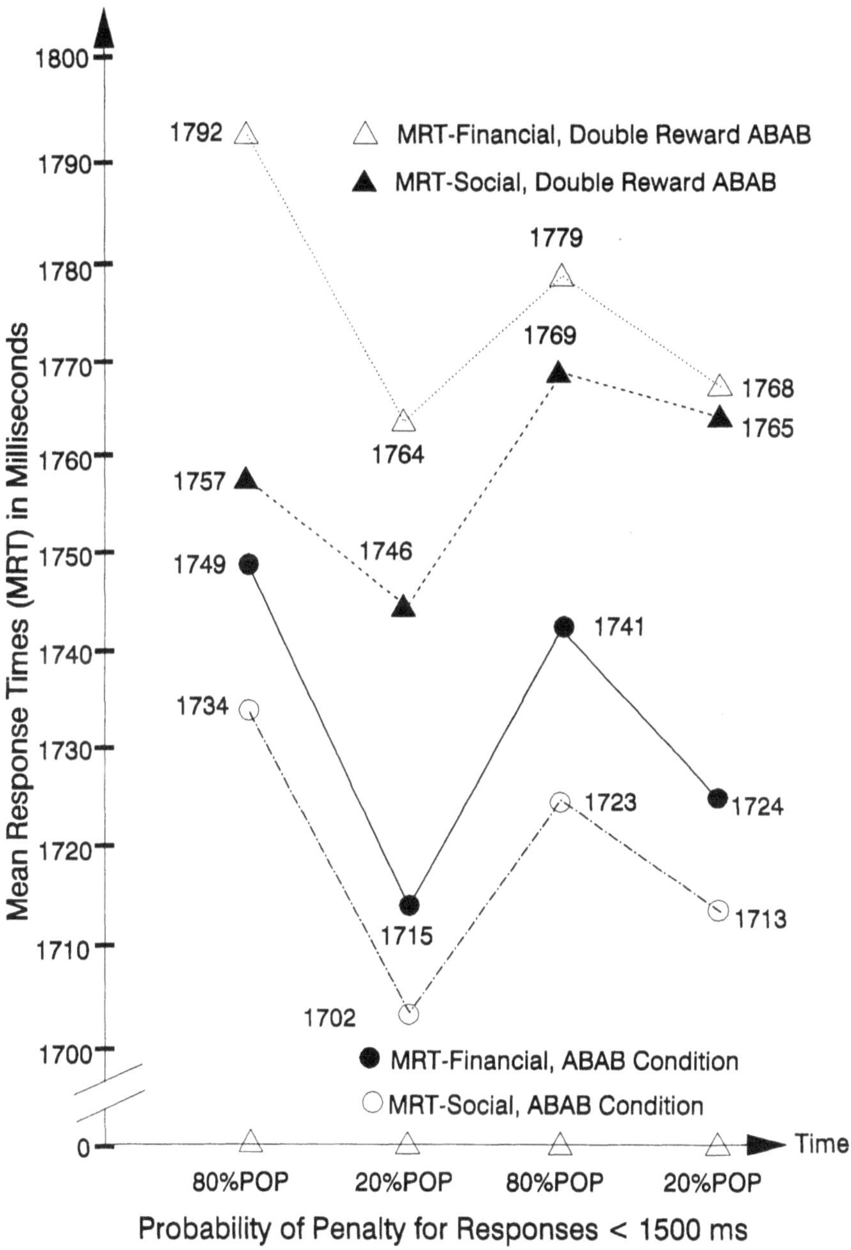

Figure 6: Mean response time shifts in the ABAB and double reward ABAB conditions. (*Figure reprinted with permission from Trimpop, 1990*).

The *Deviation from Optimality (DFO)* scores showed the predicted downward trend approaching an optimal zero deviation level (see Figures 7 and 8). The findings support both Wilde's (1982) risk homeostasis theory as well as Trimpop's (1989) risk motivation theory.

Furthermore, the data also support the notion that a change in the risk-situation influences the risk-strategy of the subjects. The findings that the DFOs responded to the withdrawal design (ABAB) and that the starting level of the financial and social condition were almost identical, both suggest that people make constantly new risk-assessments and set their target level of risk accordingly.

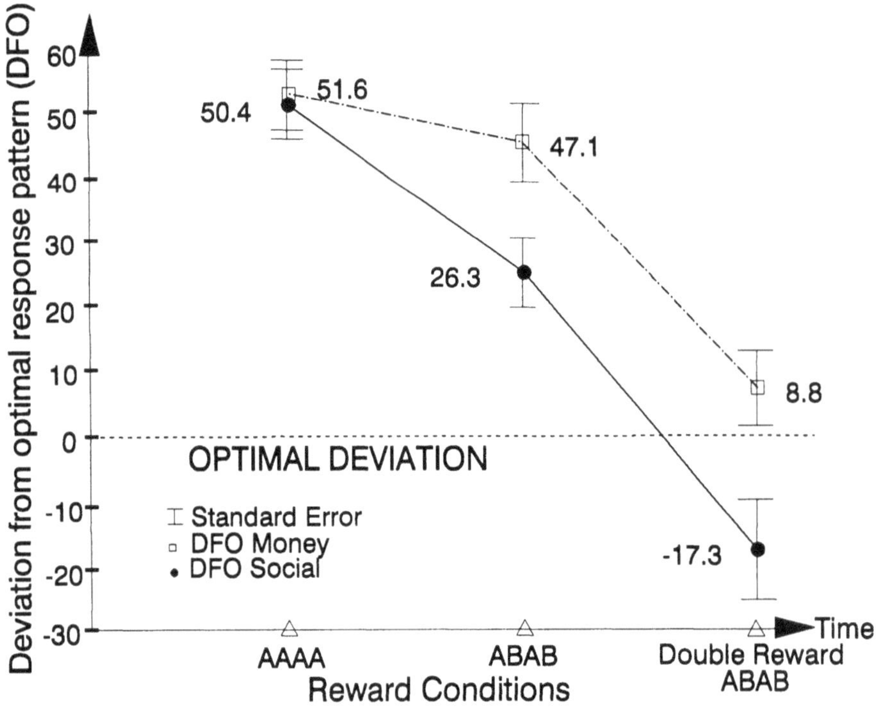

Figure 7: Deviation from Optimality (DFO) for financial and social rewards pooled across varying probabilities of penalty.
(Figure is reprinted with permission from Trimpop, 1990).

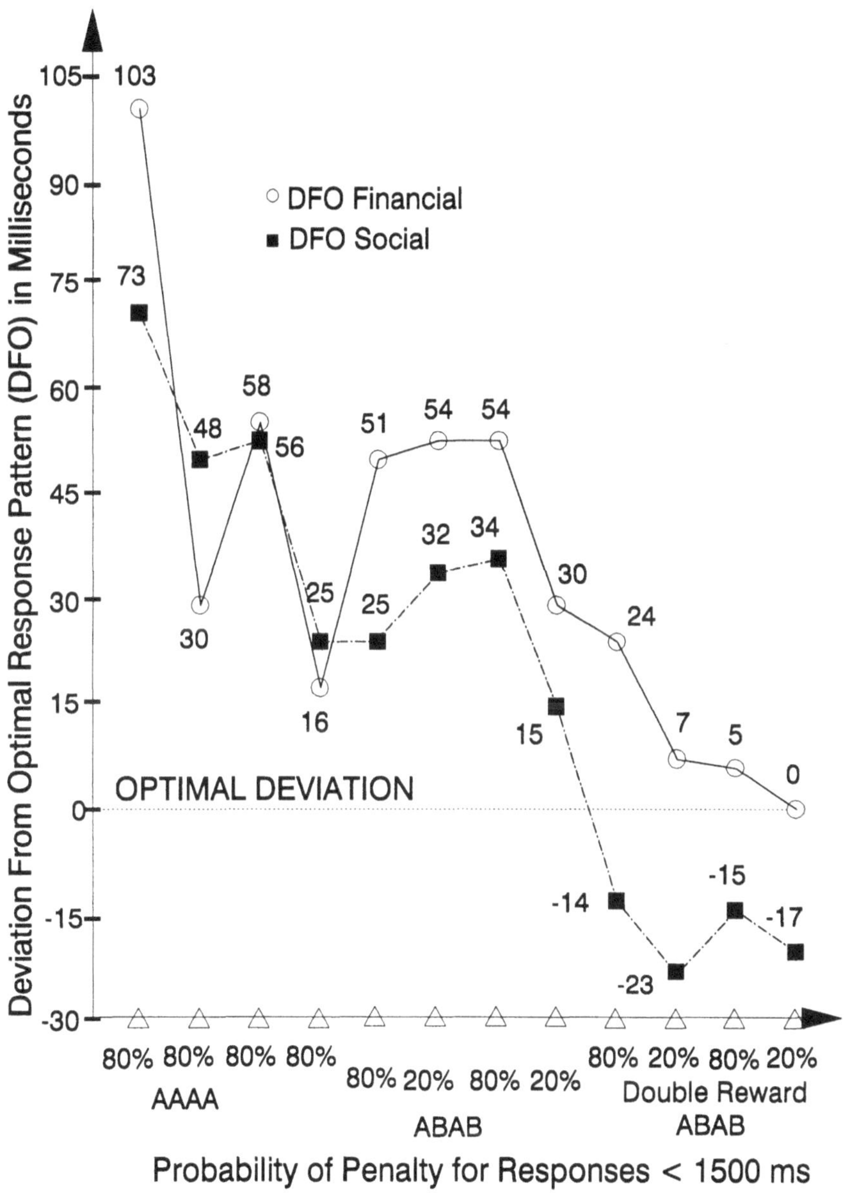

Figure 8: Deviation from Optimality (DFO) for financial and social rewards across varying probability of penalty conditions.
(Figure is reprinted with permission from Trimpop, 1990).

Physiological data also supported the assumptions made by risk motivation theory. Higher risks elicited higher arousal, as indicated by the response to changes in POP. Even the variables that did not reveal significant trends showed mean-shifts in the predicted direction, thus indicating that a more precise method of measurement, or higher levels of risk may support assumptions made by risk motivation theory even more strongly. Similar conclusions hold true for the emotional and personal experience questions. People reported emotional and personal involvement changes, such as higher perceived risks under higher POPs or in the double reward condition. The results also showed that social rewards influenced behaviors more strongly towards safe, risk averse behavior than financial rewards did.

The data of the psychomotor experiment thus offered strong support for the notion that situational changes in riskiness elicit corresponding emotional and physiological responses, as well as compensating behavior aimed at reaching a subjectively optimal level of risk taking. The concept of a risk taking personality did receive some support as subjects took similar risks across reward condition.

9.2.4 The Cognitive "Trivial-Pursuit Type" Experiment

In the cognitive task the questions were presented individually on the computer screen. They had the form of: Is London closer to Moscow than to New York? Subjects gave answers by hitting the "Y" (Yes) or "N" (No) key. Then they were asked to rate the degree of confidence in their answers, ranging from 0.5, indicating pure guessing, to 1.0, indicating total confidence. For correct answers (50% "Yes") subjects received 5 points. Incorrect responses earned zero points, if the subjects answered "No", while the correct answer would have been "Yes". If subjects answered "Yes" while the correct answer would have been "No" they received either 0 points or -5 points with probabilities of 20% or 80% (see Figure 9).

Subject Answer \ *Correct Response to Question*	YES	NO
YES	+ 5 Points	0 Points
NO	-5 or 0 Points in 80% or 20% of incorrect Yes answers	+ 5 Points

Figure 9: Payoff-Matrix for Cognitive "Trivial-Pursuit" Experiment
(Figure is reprinted with permission from Trimpop, 1990).

Heart rate, GSR and heart rate difference measures were obtained as described in the psychomotor experiment. There were four blocks of ten questions each, presented in a withdrawal (ABAB) design. The first block of trials had a probability of receiving -5 points in 80% (A) of incorrect "Yes" answers, the second block in 20% (B), then again 80% (A), then 20% (B). Each point was either added for the newspaper ranking list (*social reward condition*) or worth four cents (*financial reward condition*). After each block of questions subjects had to answer the emotional questionnaire, and seven questions asking for risk and arousal ratings in the last block of trials.

After having given their answers and certainty ratings subjects received feedback on the screen about correctness, and their individual points gained in the specific trial, as well as about the cumulative score under that reward condition. In the subject's data file, the individual "Yes", or "No" answers, the payoff, the degree of confidence, GSR and heart rate were recorded, as well as the numerical answers to the emotional questionnaires. To measure risk inclination, the variable *"Yesbias"* was developed. This variable indicates the tendency (bias) of the subject to choose the more risky "Yes" response. Yesbias was calculated by dividing the incorrect "Yes" answers by the total number of incorrect answers. This resulted in a proportion of Yes answers and the higher this proportion was, the more the subject tended to choose the more risky Yes-response. Yesbias is thus an indication of the subjects' risk strategies, while their individual skill is included as a constant part of the proportion of incorrect answers. The amount of skill was expressed as the division of all correct answers by all answers.

Following Wright`s (1984) suggestion, as a measure of *Overconfidence*, subjects' actual proportion of correct answers was subtracted from their average reported degree of confidence in the correctness of their answers (*Meanconfidence*). A positive score indicates that subject's were more certain in their answers than they should have been, given their actual scores. A negative score would indicate underconfidence in the answers.

Results of the Cognitive "Trivial-Pursuit Type" Experiment

The results strongly support the notion that subjects compensate with their behavior for the amount of risk they perceive. The very strong fluctuation of the variable Yesbias in direct response to the changes in probability of penalty provides a strong argument in favour of risk motivation theory. The hypotheses that physiological and emotional response patterns would change in response to changes in probability of penalty were also supported in the predicted directions.

The variables *Meanconfidence* and *Overconfidence* supported the assumptions made by risk motivation theory. Subjects showed significantly higher confidence ratings under the 20% probability of penalty condition than under 80% (see Figure 10). This behavior could be explained such that subjects felt more secure under the low risk condition than they actually were.

Some of the physiological variables failed to provide significant results which may be due to the amount of risk experienced not being strong enough to cause salient physiological responses. Subjects gave an average rating of 5.0 out of 9 possible points for the amount of risk experienced. Personality concepts received moderate support, as subjects who took higher risks under one reward condition also took higher risks under another.

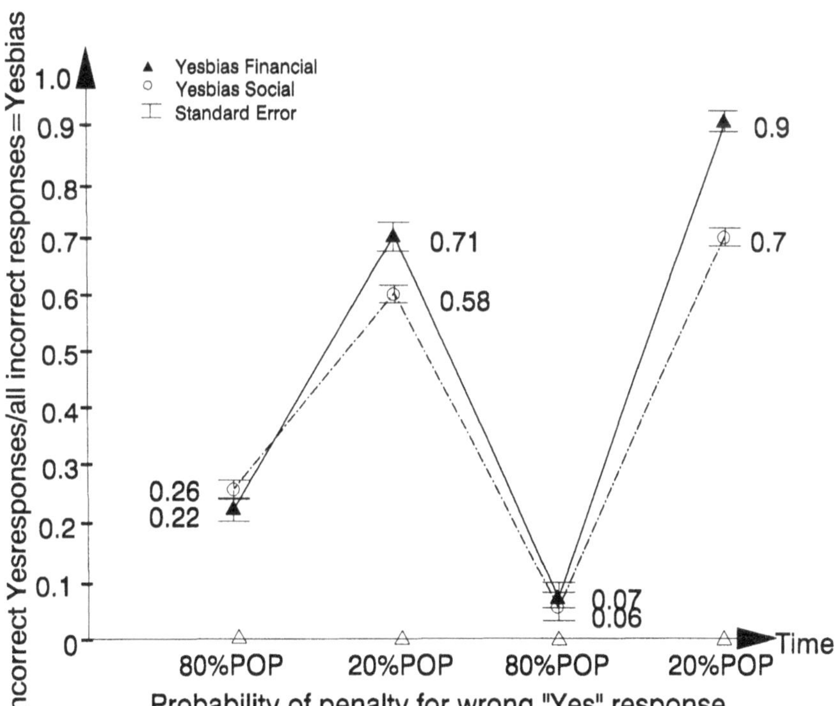

Figure 10: Tendency toward the more risky incorrect "Yes" response in relation to changing probabilities of penalty (POP)
(Figure is reprinted with permission from Trimpop, 1990).

9.2.5 Results Regarding Physiological and Emotional Measures

Physiological Measures

In relation to changes of physiological measures as a correlate of perceived riskiness strong support was found. Heart rate, GSR, and heart rate difference showed significant trends in response to changes on the probability of penalty. Several researchers (Eysenck, 1967; Zuckerman, 1979) postulated a close connection between risk taking and its physiological correlates. Supporting this assumption thus also lends support to the tie between risk taking behavior and its physiological correlates. However, there are still some inconsistencies in the results, as not all variables followed the exact pattern that was to be elicited by the ABAB design. Individual differences in responsivity may again be suggested as a possible reason for these findings. Furthermore, the type of measurements and instruments may not have been sensitive enough to pick up even small changes.

The physiological measures were also examined to test personality concepts, such as "strength of the nervous system" theories (Pavlov, 1927; Strelau, 1982; Zuckerman, 1979). These assume that risk seekers have a stronger nervous system and thus experience less arousal. It did receive only limited statistically significant support. The variables DFO, Yesbias, being a motorcycle rider, perceiving oneself to be an economic or social risk taker, having encountered life-threatening situations or being involved in high-risk hobbies all show the expected predominantly negative correlations with physiological arousal measures. Only some correlations reached a level of statistical significance. However, the consistency with which the significant and even the non-significant correlations showed the predicted negative correlation suggests that the general concept might receive stronger support if the relationship were examined differently in future studies.

Neiss (1990) and Cacioppo and Tassinary (1990) pointed out that studies relating arousal and stress or performance often show conflicting results. Cacioppo and Tassinary (1990) even stated that the research seems to suggest that arousal measures form a completely incoherent pattern, since arousal can be outcome, cause, or side-effect of psychological changes. They offered a model that may make it possible to differentiate between these different roles of arousal.

Emotional Measures

The assumption that experimental conditions with fluctuating levels in riskiness elicit corresponding fluctuations in emotional responses received strong support. On the other hand, the measures aimed at testing emotional correlates of risk taking, such as postulated in the theory of affective balance (Grossberg & Guttowski, 1987), which assumes that lower arousal levels correspond to less strongly experienced emotions as a personality construct received only little support. Again, most correlations did show the predicted direction, but did not reach a significant level. Furthermore, although Thayer's scale measured emotional activation, it does not assess other affective experiences, such as pleasure, anger, or joy.

The importance of emotional experiences related to risk taking, both pleasurable excitement and negative anxiety, requires further attention in studies testing general risk taking behavior (Izard, 1984; Larsen & Diener, 1987) and in testing risk motivation theory specifically. The significant fluctuations of the personal experience questions regarding emotions after a loss of points show that emotional experience did play a role. A spontaneous assessment of pleasure and arousal may be able to avoid an interference of complex cognitive evaluations that are also postulated to influence risk taking behavior (Lazarus, 1984; Zajonc, 1984).

9.2.6 Risk Personality and Risk History Assessment

The same 120 males described above were asked to fill out five standard personality questionnaires. The questionnaire about risk taking history was specifically designed for this study. Finally, subjects had to answer a question referring to their willingness to participate under conditions of an electric shock as a penalty in the computer experiments.

These questionnaires were used:

1) Sensation Seeking Scale (Zuckerman, 1979; 5 factors);
2) Telic Dominance Scale (Apter, 1982; 4 factors);
3) Desire for Control Scale (Burger & Cooper, 1979; 5 factors);
4) Driving Internality/Externality Scale (Montag & Comrey 1987)
5) High-Risk Personality Inventory (Keinan et al., 1984);
6) Risk History and Risky Activities Questionnaire

The questionnaire data were correlated with experimental risk measures. Due to the large number of variables only specific comparisons were made. It should be noted, that a correlation coefficient of r=.20 or more with 120 subjects has at least a power of 0.60 (Keppel, 1982) and may thus contribute some information worthwhile considering.

The risk history questionnaire was item analyzed as to the self perception of subjects in relation to being risk-inclined versus risk-averse, in relation to demographic variables, such as age, education, and occupation, as well as to their traffic and sport accident history, and in relation to their hobbies. The correlation coefficients with experimental variables and other questionnaire scores are given in Tables 2 and 3.

Table 2: Correlations of Traffic and Risk Taking Variables

* = p<.05 ** = p<.01	**Accident**	**Speed**	**Moving violation**	**Drink-Drive**	**Motor-cycle**	**Electric Shock**	**Risky Hobby**
DFO Money	-.08	-.05	.20*	.01	.14	.01	.09
DFO Social	-.06	-.24**	.14	-.10	.02	.01	-.03
Yesbias -M	-.16	-.08	.19*	.12	.01	.04	.20*
Yesbias -S	.05	.04	.13	.25**	.07	.05	.04
Moneybet	.06	.05	.24**	.06	.03	.09	-.02
Confident-M	-.12	-.40**	-.05	-.05	.06	.05	.04
Confident-S	-.07	-.19*	-.04	-.06	-.03	.06	.05
GSR-Psmot	-.02	.05	.03	-.05	-.20*	.03	-.10
HRD-Psmot	-.08	.21*	.19*	-.19*	-.09	-.03	-.22*
HRD-Cog	-.15	.19	.11	-.03	-.23**	-.04	-.21*
Lifethreat	.03	.03	.04	.31**	.29**	.10	.71**
High-Risk	.12	.21*	.19*	.10	.13	.19*	.03
Telic Scale	-.11	-.31**	-.16	-.19*	-.06	-.10	.07
SSS-Total	.15	.23**	.25**	.30	.12	.28**	.10
TAS	.13	.23**	.25**	.13	.26**	.25**	.19*
ES	.10	.19	.11	.28**	.10	.20*	.12
DIS	.01	.23	.27**	.26**	-.09	.22*	.08
ControlTotal	-.14	-.01	-.07	-.25**	-.10	.00	-.08
Controlwish	-.12	-.03	-.03	-.21*	-.09	.01	.19*
Planning	-.07	-.30**	-.12	-.21*	.01	.02	.10
Seriousness	-.11	-.25**	-.07	-.14	-.02	.05	.09
Leadership	-.11	.13	.01	-.21*	-.03	.04	.07
Preparation	-.12	-.14	-.16	-.19*	-.02	.09	.21*
Violations	.44**	.20*	1.0	.30**	.14	.06	.18*
Drink-Drive	.20	.29**	.29**	1.0	.23**	.03	.20*
Motorcycle	.04	-.19*	-.22*	.23**	1.0	.03	.27**
Speed	.20*	.07	.29**	.29**	.07	.16	-.04

Table 3: Correlations of the Experimental Risk Taking Variables

* = p<.05 ** = p<.01	**DFOMoney**	**DFO Social**	**Yesbias M**	**Yesbias S**	**Moneybet**
DFO Money	1.0				
DFO Social	.72**	1.0			
Yesbias -M	.24**	.19*	1.0		
Yesbias -S	.26**	.11	.34**	1.0	
Moneybet	.14	.13	.13	.06	1.0
Risky Hobby	.09	-.03	.21*	.04	-.02
Confident-M	.10	.19*	.31**	.02	.08
Confident-S	.23**	.21*	.06	.17	.05
Telic Scale	-.17	.03	.06	-.05	-.14
Arousalavoid	-.33**	-.08	-.02	-.14	-.02
Leadership	.23**	.02	.08	.01	.09
Planning	-.13	.05	.00	-.06	-.25**
High-Risk	.37**	.11	.01	.13	.16
SSS-Total	.23**	.07	.04	.22*	.11
TAS	.20*	-.05	.03	.20*	.15
ES	.12	.01	.04	.21*	-.14
DIS	.31**	.17	.07	.30**	.10
Violations	.20	.14	.18*	.13	.23**
Drink-Drive	.00	-.10	.12	.25**	.06

The assumption that experimental risk taking behaviors are connected to personality traits and to other types of risk taking activities did find significant empirical support. Risk taking behavior variables showed positive intercorrelations, many of them reaching significant levels (see Table 3). Across different tasks, however, somewhat different types of risk taking behavior seemed to have been active. The amount of money bet in the dice-rolling task did not correlate significantly with the other tasks.

The correlations of the risk taking variables (Yesbias, Overconfidence, DFO, Mean Response Time) between cognitive and psychomotor tasks did not exceed $r=0.23$, $p<.01$. Although these positive correlations are significant and in the predicted direction, they only account for less than 5% of the variance explained. This leaves 95% of the components involved in the relation between different risk taking tasks unexplained. This finding can be seen as a strong indication for the multifacetedness of risk taking. If the correlations had been very high it could have been inferred that the tasks measured the same types of risk. This way, however, the concept of risk taking behavior consists of a small but significant general component, while it varies between situations very considerably.

Within the psychomotor-task, under the social and financial reward condition the risk taking variables (DFO, Mean Response Time were intercorrelated at a level of at least $r=.70$ ($p<.001$, see Table 3). In the cognitive task, the Yesbias financial and Yesbias social variables were correlated at $r=.34$, and Overconfidence measures at $r=.47$ (both $p<.001$). These results indicate that risk taking within the same type of behavior is somewhat more stable, but across different rewards 50% or more of the variance explaining the relationship still remains unaccounted for.

These findings suggest, that risk taking behavior possesses a component that generalizes across different reward patterns and across different types of tasks, but still leaves a lot of room for individual and situational differences (Zuckerman, 1979). Large individual differences between subjects and the very different nature of the risk taking tasks, namely betting money, answering "trivial-pursuit" type questions and estimating response times does not make this result a surprising one.

Also, experimental and personality questionnaire assessments were significantly related to each other. The *Sensation Seeking Scales*, the telic dominance subscale *Arousal-Avoidance*, and the *High-Risk Inventory* all showed significant correlations with experimental risk taking behavior in the expected direction (see Tables 2, 3). Furthermore, both the self-report and the experimental measures showed the predicted significant correlations with risk taking in traffic, with participating in risky activities and with considering oneself to be a social, financial, or physical risk taker.

Especially risk taking in the social reward condition showed significant correlations with questionnaire measures, and with the reported acts of drinking and driving, higher speed and more moving violations (Table 2). Surprisingly, no significant relation between any of the experimental measures and being involved in traffic accidents emerged. It could be argued that drinking and driving, as well as receiving a moving violation ticket are more social and financial risks than a physical one. Surprisingly weak ($r=.20$, $p<.05$) correlations between drinking and driving and traffic accidents also suggest that traffic accidents may not be caused predominantly by risk seeking, but rather by risk-tolerating behavior. Also, a lack of psychomotor abilities, lower attention span, sleepiness or, most importantly the expectation of being drunk may be influential instead. Being drunk could serve as a socially acceptable excuse - maybe even as a requested behavior pattern - for being careless and reckless.

Consistently negative correlations between *Desire for Control Scales* and drunk driving were found (Table 2) and may be explained by the fact that alcohol consumption reduces psychomotor control (Wilde, Trimpop & Joly, 1989), as well as judgemental skills. It is thus not surprising that people high in the desire to maintain control refrain from drunk driving.

One problem with the findings reported so far is that the real life activities and accident histories are based upon self-reports. It would be interesting to examine and observe certain high-risk occupational groups, such as fire-fighters, stuntmen, and policepersons, or high-risk sport groups, such as race-drivers, divers, or gamblers, to confirm these findings. Overall, however, the assumption that laboratory risk taking behavior is related to real-life aspects of risk taking, such as to participate in risky activities, to take higher risks in traffic, or to consider oneself a risk taker, was supported.

Due to ethical considerations physical risks could not be examined the same way as financial and social risk taking. Arthur (1986), however, argued that the expectation of a physical punishment, such as an electric-shock is as stressful as the actual delivery of it. Thus, the willingness to endure a physical penalty was used to measure physical risk taking, but it did not show any significant relationship with experimental risk taking measures. However, significant correlations (Table 3) were found with all five *Sensation Seeking Scales* and the *High-Risk Personality Inventory*.

The relation between risk taking and aspects of perceived personal control became also apparent in the verbal response some subjects gave to the electric-shock question. Some said they would participate, but be much more careful to avoid penalties, while others said that they would do it since they could control whether they would receive a shock or not by either answering "NO" in the cognitive experiment or by making slow responses in the psychomotor experiment. Others said: "Sure why not?" Yet, others answered: "Of course not, why would anybody?" A structured analysis of reasons for participating in such a study could provide interesting insights into risk motivations of individual subjects. Future studies could incorporate a physical penalty, such as an electric-shock, or a noise-blast, and compare it to a physical reward, such as food.

Although subjects generally showed positively correlated risk taking behavior across experimental manipulations, these correlations only explained between 3.5% and 10% of the variance. This result gives only limited support for the concept of a *"general risk taking personality"*.

In comparison to the strong situational factors, personality factors are far less influential. That significant relationships emerged across all different domains, indicates that there may be a general risk seeking trend. However, the low amount of variance explained by this personality factor and also the missing relation to accidents puts doubt on the effectiveness of selecting general risk taking personalities to reduce accidents.

Future studies should try to examine the complex cognitive processes in greater detail, assess emotional (e.g., Affect Grid, Russel et al., 1989) and physiological correlates (e.g., EEG, EMG) of risk taking behavior more directly and spontaneously. Then it would be important to test whether the causal assumptions and the latent structures in risk motivation theory, such as *Target Levels* and *Motivation,* can be confirmed in structural linear equation models, such as performed by LISREL, or EZPATH. Such an analysis could strengthen the support for risk motivation theory.

To summarize, the studies described above offered substantial support for the claim of *Risk Motivation Theory* to explain risk taking behavior holistically. Yet, its general validity still has to be tested in relation to women, other cultures, and other populations within the society.

Chapter 10: What Does Risk Motivation Mean for Our Daily Lives as Researchers and as "Normal" Risk Takers ?

In the final chapter, I shall summarize which answers have been found to the questions about risk taking behavior in the introductory chapter. The conclusions drawn from the studies described in the preceding chapters will be discussed in relation to research on risk taking and in relation to the real-life of "normal" risk takers, and researchers alike, as oftentimes they overlap. Some personal conclusions will end the chapter and the book. These were the questions asked about risk taking behavior:

1) What is risk taking behavior and how do we perceive risks?

2) What personality type engages in risk taking behavior? And is it mainly determined genetically or by situational characteristics?

3) What drives people to take risks, and what do they feel before, during and after the performance?

4) How do we control and master possible negative consequences of risk taking behavior and how do we compensate for different levels of perceived risks with our behavior ?

5) Is there a connection between different types of risk taking (financial, social, physical) and can we explain it in one "holistic" model ?

6) What does ***risk motivation*** mean for our daily life and how can we modify it?

10.1 Research Implications of a Motivation for Risks

To answer these questions, I have reviewed the literature, conducted the studies described in chapter 9 and found that many approaches offer part of the solution to understand what risk taking behavior is.

10.1.1 Is There a Personality Factor in Risk Taking ?

The influence of individual differences on risk taking behavior was clearly evident in the literature. The sample did not consist of professional risk takers, such as stuntpersons, fire-fighters, or race-car drivers, but mainly consisted of young white male citizens and university students of a medium sized Canadian city, who considered themselves average in their risk seeking tendency, and who ranked average in the risk history and activities scale. Nevertheless, finding the predicted significant positive correlations between personality trait, experimental, and real-life risk taking variables, did support the notion of a general risk taking personality trait. However, the assumption of a *very strong* personality trait influence on risk taking behavior (Eysenck & Eysenck, 1976; Strelau, 1982; Zuckerman, 1979) was not supported. Evidence for a general risk taking trait, although statistically significant, was not as strong as these theories predicted. This could be due to the sample, but given that most studies and theories were based on similar samples, the findings of this study raise doubt about the level of influence personality traits exercise on risk taking.

Although part of the literature found support in favour of genetic and evolutionary components of risk taking (Bouchard, 1986; Buss, 1988; Darwin, 1859/1968; Dawkins, 1976 see chapter 3), these aspects can generally not be tested experimentally. Finding significant personality components across a large variety of activities and types of rewards, however, could be viewed as offering some support for the correctness of such an assumption. Future research could examine risk taking tendencies in experiments, real-life and questionnaires among monozygotic and dizygotic twins reared together and apart, or in adoption studies. To construct a fair test for the effects of the heritability of individual differences in risk taking tendencies would require a truly longitudinal study. Maybe some inferences can be drawn from primates, or other animals where controlled mating is easier, quicker and less unethical. Overall, the personality risk factors contribute some significant information to increase the understanding of risk taking behavior and the nature/nurture debate.

Buss (1989) stated that asking whether personality traits or experimental manipulations control more variance is futile, as the designs of experiments are usually biased in favour of either one. Cooper and Richardson (1986) argued that studies comparing different phenomena have to make sure that the comparisons are based on equally strong operationalizations, manipulations and measures. Otherwise, finding a difference in explained variance represents a methodological artefact, rather than a true difference. Nonetheless, the correlation coefficients of personality variables are predominantly in the range of r=0.20-0.40 (Buss, 1989) throughout the literature and in the study described in chapter 9, inspite of acceptable reliabilities and extensive testing in numerous studies.

I conclude therefore that personality risk factors play a significant role in risk taking behavior, but that they play a less important role (5%-25% of variance explained) than situational factors do. Unequal measurements, however, do not allow to ascribe exact figures to the variance explained by either one. Yet, the results do have important implications for research on risk taking (e.g. personality traits as a covariate) and safety measures (e.g., task-specific selection and training of risk takers).

10.1.2 What Is Our Motive to Seek Out Risks ?

There are many different conceptualizations of intrinsic and extrinsic motivation (Deci & Ryan, 1988; Heckhausen, 1980). A common element in most of these approaches is that there are pleasurable experiences connected to the *act of performing some activity*, rather than with reaching a goal (Heckhausen, 1980). These pleasurable events can be rewarded physiologically, emotionally or cognitively. The same stimulus (e.g., money), can sometimes influence extrinsic motivation, (e.g., in giving a salary), or intrinsic motivation, as in games (Kruglanski et al., 1975).

Most theories of expected utility (e.g., *SEU theory* (Vroom, 1964); *portfolio theory* (Coombs, 1972); *conflict theory* (Janis & Mann, 1977); *prospect theory* (Kahneman and Tversky, 1973)), have so far not considered emotional, physiological components, or the positive aspects of risk taking, or the importance of non-material rewards. The same criticism holds true for many theories of risk taking behavior in traffic, such as the *zero-risk-model* (Summala, 1988), or the *threat-avoidance model* (Fuller, 1984). Even *risk homeostasis theory* (Wilde, 1982) considers risk taking to be an extrinsically motivated activity with risks being tolerated. Risk homeostasis theory does not account for risk as a desirable, pleasant activity, which is actively sought out by people and offers intrinsic and extrinsic rewards, such as pleasant emotions or pleasant arousal-highs. The literature on emotion, motivation, and the findings of the studies in chapter 9, however, suggest that these factors might be even more effective in addressing motives for risky or safe behavior than material rewards, arrived at by merely goal-oriented behavior.

In the studies described in chapter 9, in every instance of the ABAB blocks the deviation from an optimal risk-strategy was on the risky side. This indicates that subjects generally test their limits, luck, and chances by taking risks. Furthermore, *no subject* chose a completely safe approach, as each subject made several too risky responses. This is further support for the postulate of an intrinsic necessity, desire, and motivation to take risks. The measures of who takes how much risk when and where, shows large individual differences, while changes in probability of penalty influenced everybody. The one significant deviation from the risk-inclination was found when a *doubled incentive for safe behavior* was introduced, which significantly altered the behavior towards a more cautious approach.

Thus, when applying the literature on motivation specifically to risk taking behavior it seems plausible to postulate that people receive intrinsic rewards for risk taking, such as hormonal pleasures (Zuckerman, 1984), physiological pleasure center stimulation (Olds & Fobes, 1977), emotional pleasures, (Solomon, 1982), the experience of slight incongruencies between desired and target level of risk (Fiske & Maddi, 1961), or the unity of action-oriented and goal-oriented risk taking behavior (Heckhausen, 1980). In other words, we take risks for the intrinsic pleasures associated with it, as well as for the extrinsic benefits, such as increased material profits.

The Power of Incentives

Risk motivation theory (Trimpop, 1989) and risk homeostasis theory (Wilde, 1982) postulate that incentives for safe behavior are a powerful motive to reduce risk taking tendencies. Whereas changes in riskiness of a situation will lead to higher perceived control and thus to compensating behavior. As a result, the externally introduced safety benefits are virtually cancelled. On the other hand, although introduced externally, incentives for safe behavior manipulate the intrinsic motivation of subjects by changing the desired, or target level of risk. The target level of risk is influenced by a higher perceived benefit of safe and higher costs of risky behavior, and thus the subject chooses a less risky strategy.

The shift from risk-inclined to safety oriented behavior was extremely strong in the Double Reward ABAB condition of the psychomotor task, described in chapter 9. As Figures 7 and 8 demonstrate, the incentive was so strong that in the social reward condition subjects even overcompensated, as indicated by negative DFOs. This effect had not been observed in previous experiments on risk homeostasis theory. This finding indicates that target levels of risk are constantly readjusted according to the subjects' perceptions of risk. Overcompensation suggests that subjects had the illusion to incur more risk than they actually did.

This finding is thus a powerful suggestion for using incentives to increase safer behavior. The finding that the effect was significantly stronger for the social reward condition than for the financial one is an indication for the differentially motivating effect of incentives. They do not have to be financial in nature. Most likely, social incentives have a higher likelihood of triggering intrinsic motivations. The data also showed that incentives rather than punishment should be used to increase safety oriented behavior. Thus, *"Fame"* oriented rewards may be more effective in increasing safety behavior than *"Fortune"*, or *"Pain"* incentives.

Another argument in support of using rewards comes from clinical research. When people develop phobias they feel rewarded for avoiding a certain situation, in spite of strong negative effects they encounter as a consequence of their avoidance behavior (Gleitman, 1986; Hilgard & Bower, 1966). For example, both the fear of being in open places (agoraphobia), or in closed places (claustrophobia) puts severe limitations on a person's social and professional life. Yet, in spite of increasing problems, such as job loss, social isolation or other penalties, phobic persons often maintain this behavior until an alternative behavior is established as more rewarding than to avoid feelings of anxiety (Fliegel et al. 1980).

Thus, I strongly suggest that instead of rewarding people for avoiding risk by not giving them a penalty, they should be motivated to perform safe behavior by giving them incentives. Whether incentives work most effectively on a physical, economic or social level, or whether they are presented contingent upon performance or in a lottery may vary greatly between individuals, situations or cultures, and would be worthwhile to investigate. It appears, however, to be essential that people perceive the incentive as such, regardless of its value to others. Whether it is material or spiritual in nature, or even illusionary, is largely irrelevant, as long as it is perceived as valuable and as an incentive for the desired behavior.

10.1.3 How Do We Master and Control Our Risk Motivation?

In chapter 6, I stated that the need to control and master a situation is just as intrinsically and extrinsically rewarding as the desire to seek out the unknown. Thus, we strive for gaining increased control and adjust our behavior accordingly, even if the perception of control is illusionary. When presented with a situation of illusionary higher uncertainty and unfamiliarity subjects in the dice-experiment (see chapter 9), bet less money, even when a presumably less unfamiliar and less "risky" situation followed. When faced with the illusion of high control and low riskiness, this illusionary security seemed to have dominated their behavior. This was true even in the situation designed to increase the perception of risk and to decrease the perception of personal control.

Virtually all subjects said that they knew that the chances of making a correct prediction of the roll being above or below ten was always 50%; yet many still engaged in some form of superstitious habits, such as rolling harder for high numbers, betting on below 10 after two to three rolls above the value of 10, and attributing their success to the kind of day they have had, their luck in love, in games in general, the time of day, etc.

It is quite conceivable that subjects were more suspicious about what happened to the die "in the box" and thus bet less money, although their rational mind told them that there is no reason to do so. The illusion of control (Langer & Roth, 1975) and the resulting behavioral reactions are a quite robust phenomenon, which may even have a very valuable function for our well being, as Taylor and Brown (1988) and Jeffery (1989) pointed out. These findings support Taylor and Brown's (1988) assumption that most people do not perceive their own circumstances realistically, but have a positively coloured illusion about themselves, their skill, and their abilities to control the environment. With control and risk taking being closely related, it is not surprising that people tend to take higher than optimal risks (e.g., generally positive DFOs. 80% are better drivers than the average), as they also believe they have high control. Taylor and Brown (1988) and Taylor et al. (1992) stressed the positive effects of an overly positive self-illusion and optimism regarding mental well-being, coping and psychological benefits for general problems and AIDS. But, of course, there has to be a point after which too much illusionary optimism and thus too high risks lead to severe drawbacks and ultimately death. Future research should examine such relations and take into account that people will likely be risk-inclined, due to illusionary optimism and overconfidence.

Such findings confirm the importance of incorporating individual perceptions of personal control, even illusionary ones, when analyzing risk taking behavior. They also show that the perception of control, or the lack of it, leads to corresponding risky or safe behavior. Thus, creating the illusion of controlling traffic accident consequences with technical features such as seat-belts, ABS, four-wheel drive, helmets, etc. will seduce people to show more risky, thoughtless or even careless behavior.

"Don't worry, I have got it all under control" is one of the famous *last* proverbs used before accidents.

10.1.4 How Do We Compensate for Perceived Risks ?

A major underlying assumption of risk homeostasis and risk motivation theory is the behavioral compensation for perceived changes in riskiness. It is postulated that people will show compensating behavior to reduce a discrepancy between their perceived and desired level of risk. Apart from correlational support in the literature (see chapter 8), this assumption was also strongly confirmed experimentally in an ABAB design (see chapter 9). It even held up across different tasks, different rewards, over time and across repeatedly changing levels of risk.

Furthermore, the risk taking strategy variable DFO (*Deviation from Optimal Response Pattern*), which partialled out individual skill factors, also showed these behavioral shifts as a consequence of changes in probability of penalty. This finding had been predicted by risk motivation theory, claiming that **each** change would produce a new perception and resetting of the target level of risk. On the other hand, risk homeostasis theory would not have assumed a sensitivity to these short-term experimental manipulations, but emphasizes instead the long-term effects on risk taking strategy.

What Differentiates RMT from RHT ?

Risk motivation theory is based upon risk homeostasis theory in the notion of a compensatory behavioral feedback loop for perceived discrepancies between desired and targeted risks. Risk homeostasis theory, however, does not account for multiple, constantly shifting target levels of risk and the risk seeking tendency of people. RMT furthers the postulates of risk homeostasis theory in that it assumes that physiological and emotional processes undergo the same compensatory functions as the rational cognitive processes do on which risk homeostasis theory largely focusses upon. It also adds the notion that people take risks for the intrinsic rewards associated with performing risky activities, and that this process also underlies a compensatory feedback loop. It is less extreme in that it does not necessarily assume that a homeostatic balance is ever achieved, as target levels are constantly changing and there are always new discrepancies. In these respects, results of the studies in chapter 9 favour risk motivation theory. DFO shifts were significantly influenced by changes in riskiness and were reset after each perceived change in the degree of riskiness, while overall trends approached an optimal level of risk taking (see also Figures 7, 8). Thus, risk motivation theory aims to expand risk homeostasis theory in:

a) *the postulation of multiple target levels of risk that require constant compensation and may never reach a homeostatically controlled optimal target level of risk,*
b) *the strong emphasis on individual motivational processes, rather than collective behavior of a larger population,*
c) *the importance of emotional and physiological aspects in the assessment and evaluation of risk taking behavior,*
d) *the addition of activity oriented rewards (desire for risk) to goal-oriented rewards (tolerance of risk).*

10.1.5 Can We Describe Risk Taking Behavior Holistically?

The major distinction of risk motivation theory from other theories is that it combines many facets of risk taking in one holistic model (See Figure 11). It accounts for the desire to take risks, as well as for the desire and the process to control it. It includes processes that are not consciously perceived by the person involved, such as physiological and elementary emotional processes. Risk motivation theory stresses emotional, physiological, and cognitive aspects of risk taking, and a behavioral compensatory feedback loop, designed to optimize the overall *benefits of taking risks*.

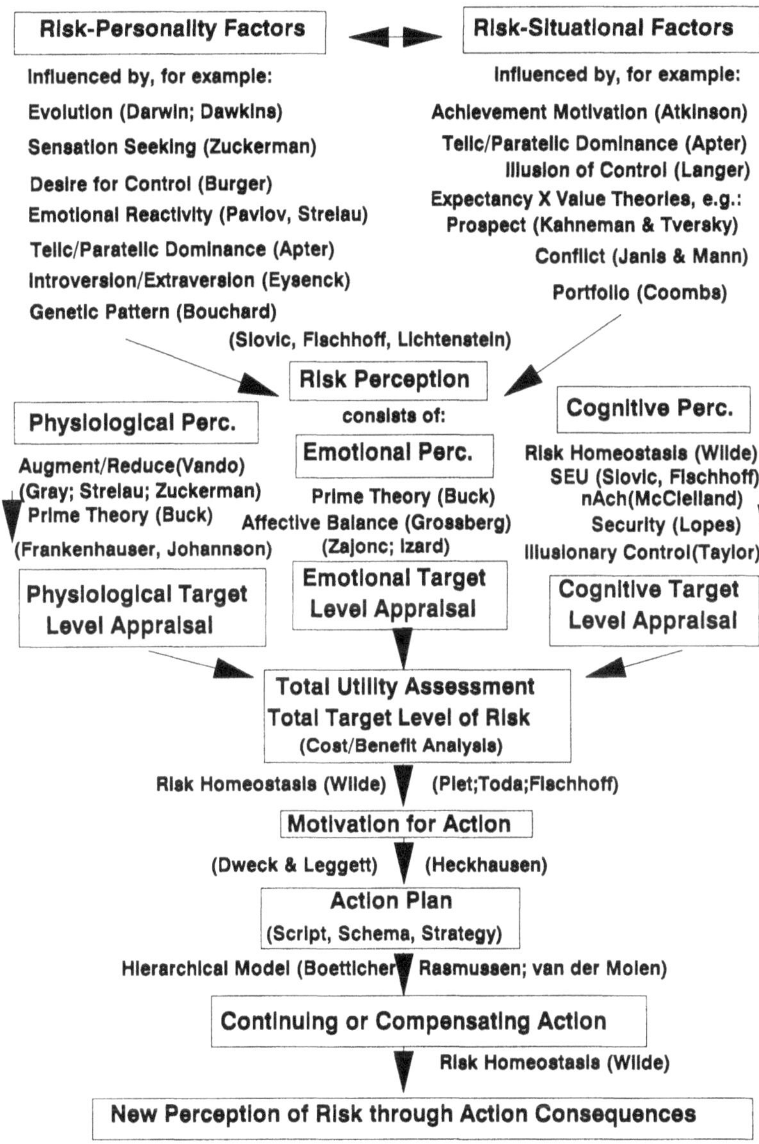

Figure 11: Theories incorporated in Risk Motivation Theory (RMT)
(Figure is reprinted with permission from Trimpop, 1990).

10.2 Implications of a Risk Motivation for the "Real Life"

The literature and studies in this book did provide strong empirical support for postulates made in risk motivation theory as well as for the theory it is derived from, namely risk homeostasis theory (Wilde, 1982; 1986). That means that people adjust their behavior according to the risk they perceived.

While this finding sounds trivial in the case of being more cautious when the danger is high, the finding also means that people will voluntarily increase their exposure to danger when the perceived risk is low! This has very important implications for industry, traffic-regulations, health issues and one's personal life. The findings essentially suggest that when we make the environment safer, people will compensate for this increased safety by taking more risks.

10.2.1 Implications for Accident Reduction

The literature gives examples for increased speed when "buckled up" (Streff & Geller, 1988), equally high numbers of accidents with anti-lock braking systems (Aschenbrenner & Biehl, 1988; 1993), increased numbers of neck and spine injuries with decreasing head injuries due to the "more efficient" use of football helmets as a battering ram (Torg et al., 1979) and many more instances of no reduced fatality as a consequence of externally introduced safety measures (Adams, 1985; Wilde, 1989).

For industrial safety it is important to keep this relationship in mind when designing a workplace. The less active input a worker has into operating machinery, the higher is the likelihood of boredom, underarousal and thus compensating risky behavior. Interviews conducted in industrial plants (e.g. Trimpop, 1989) certainly point to risk increasing behaviors, such as duelling with rivets, oil cans, running races with electric cranes or fork lifts.

Therefore, it seems to be important to keep the workers actively involved in the design and execution of their jobs. The amount of responsibility experienced is probably proportional to the amount of risk experienced. Unless workers are replaced by robots their risk taking and arousal desires have to be considered if accidents are to be avoided.

The research portrayed in this book also supports Wilde's (1982; 1986) and Trimpop's (1989) postulate that one way of interrupting the vicious circle of risk and safety compensation behavior lies in giving incentives for safe behavior, as incentives are assumed to change risk motivation. According to de Charms (1968) and Deci (1975) incentives may increase or decrease intrinsic motivation, depending on how they are perceived. For example, giving people material rewards for something they enjoy doing has been found to reduce their desire to perform that activity (De Charms, 1968; Heckhausen, 1980).

Thus, it is important to determine in each situation and with each person whether the reward offered changes an intrinsically oriented into an extrinsically oriented one (e.g., money for sex) or whether it adds to the intrinsically rewarded desires and activities (e.g., free condoms to reduce the danger of AIDS). In the study described in chapter 9, offering doubled rewards changed the risk motivation and led to safer behavior, even to behavior that was more safe than optimal, against the criterion of maximizing scores.

Although incentive programs (e.g., Fox et al., 1987) did show some long-term effects, it may be expected that due to habituation, people will get used to these increased rewards, and thus, after a while, take more risks again. It would, therefore, be desirable to change the kind of rewards periodically, make the rewards flexible to accommodate individual needs and differences, and make incentives non-material, as well as material in nature. Non-material rewards, such as praise and fame are often more congruent with an intrinsic motivation than material ones.

Accident countermeasures can be introduced and controlled by the government; they can also be utilized effectively by industry, and perhaps most importantly - in the family. If we succeed in teaching and rewarding people from an early age on for showing the desired risk-conscious behavior, it would be much more effective than if we try to change the habits of an older person. Similarly, training workers in perceiving risks and reducing those that are undesired, promises to be more effective when started right at the beginning of their employment.

Another important aspect influencing risk taking is risk perception. If a risk is not perceived as a risk it cannot be responded to. Thus, if workers, drivers, or other people are not taught in recognizing dangers, accidents are bound to happen. For example, usually the right hand brake of a bicycle works on the rear wheel, while that of a motorcycle works on the front wheel. Thus, if a rider switches vehicles without knowing the difference, applying equal pressure can have drastically different results. In a more complex setting, such as an industrial plant, there are many hidden risks, such as fumes, broken safety devices, or new and unfamiliar equipment. Examining and pointing out these risks and involving workers in discovering and understanding, as well as coping with these risks would provide an excellent opportunity to make risks visual, and thus increase safety conscious behavior in respect to these risks.

10.2.2 The Insurmountable Desire for Challenges in Our Daily Lives

But: The results of this book also indicate, that no matter what the rewards and penalties are, most people will still take risks. The literature showed that risk taking is essential for survival, it is fun, and it is rewarded intrinsically, as well as by society. Thus, we should not even try to **eliminate** the risk taking behavior of either young or older people. Instead, we should rather channel it into domains that are at least as risky and enjoyable for the individual, but less risky for society, or members of society who are merely tolerating that kind of risk. For example, it is much cheaper for a society if a race-car driver makes an error, than if the driver of a truck with toxic waste material does. If the skydiver overestimates his/her skills and opens the parachute too late, or not at all, the consequences are severe for him/her, but by overestimating his/her skills, an airline pilot produces a much more dreadful risk for society. If a motorcycle driver decides to try whether he/she can drive on the back wheel only, the consequences may be different if the attempt is made on an empty backyard or in downtown traffic. If a manager decides to sit back and read the newspaper at the workplace the consequences could be different than if the assembly line worker decides to do the same.

The point I want to emphasize is that we have to try to make people actively choose which risks they want to take and to limit the situations they take them in, as well as seek out other situations or locations where and when to take risks. People should be trained in weighing the costs and benefits of this action. If the 18 year old decides that it is more important to impress the "love of his/her life" by doing a "wheely" in downtown traffic than to do it in the backyard, then this behavior should come about as a decision, not as an over-representation of feelings of inferiority. On the other hand, if the business person decides to replace the little store on the corner with a more profitable business, then he/she should emphasize and react strongly to the emotional and social risks for his/her self-worth and that of the other people involved, rather than letting the cognitive-material aspects be over-represented in the total utility assessment.

To illustrate further that risks can only be channelled and not abolished, I have yet to meet people who do not get a glow of pride in their eyes when talking about risky experiences they mastered. Whether it is getting lost on a vacation trip and finding the way, or coming out of an almost lost sports game by making an impossible shot, or pulling off from the edge of bankruptcy. Even stronger, the areas of social risk taking confront us daily with decisions where we have to take risks, and where we see compensating behavior at work. The friendly boss will be spoken to much more openly than the bad-tempered one. Inviting a long-term relationship partner for dinner is usually perceived as less risky (there are of course situational exceptions) than going on that all important first date. Choice of words, clothing and restaurant will be adjusted accordingly. Yet, we all have to take these and other risks and we even enjoy them, up to a certain degree.

Anticipatory joy (going on a roller-coaster), anticipatory fear (for example in suspenseful movies), as well as the actual experience of emotions and physiological arousals are very intense under risky conditions. The increased pleasure from this intensity provides a powerful incentive for searching out situations that can give us this feeling again.

How much risk and intensity feels enjoyable is, however, subject to many situational and individual differences. Listening to interviews with race-car drivers, such as Emmerson Fittipaldi, or talking to stuntpersons (Piët, 1987), or listening to myself after riding the motorcycle rather quickly seems to suggest that the higher the experienced risk is, the higher is the amount of pleasurable arousal, and the higher is the amount of energy spent on controlling the dangers. Driving at 200km/h on the Autobahn raises the alertness level and the perception, and thus the anticipatory control to compensate for unexpected events, such as driving errors of other drivers. The amount of thinking, testing, protection and money that is spent to control the dangers for public figures, or for drivers on the race tracks is probably somewhat proportional to the amount of risk for the individuals involved. The amount of effort and safety checks for manned spaceships are by far larger than the ones for unmanned ones, especially after accidents. As high risk-takers seem to increase their effort to control the risk they take, we may understand the processes involved in "normal" risk taking better when studying "extreme" risk taking. What this balance of control and risks also means, is that increased subjective risk taking may not at all lead to increased accidents, but could be compensated for by increased attention, control and mastery.

The effects of illusionary control also have potentially important implications. As has been pointed out by Langer (1975), Burger and Cooper (1979), and Taylor and Brown (1988) illusionary control, optimism and skill are powerful phenomena prevalent in most of us. Since this book suggests that we are all risk-takers to the level that we feel comfortable with, and as we are also presumably overestimating this level, we should on average take consistently too high risks! This would have very important implications for our survival, namely that unless another factor compensates for this phenomenon, we should have either become extinct or should eventually do so.

It does not seem wise to wait to the last split-second of human life on earth to see the results of the "field observation" of whether too high risks due to illusionary optimism ultimately leads to extinction. Thus, the relation between illusionary control, optimism and risk taking behavior of individuals and populations alike, should be studied in controlled experiments.

Finally, further research has to show whether the initially supportive findings for risk motivation theory are due to "illusionary optimism" of the author, or wheter RMT is an important contribution to the fascinating field of human feelings, thought and behaviors, named psychology.

10.3 Personal Conclusions

What motivates me to take risks? Why do I enjoy motorcycling, playing games, parachuting, giving a presentation, courting a woman, travelling to foreign countries, discussing controversial issues openly and vehemently, and other risk taking activities? To come closer to an answer to these questions was the reason for me to start this book. After 36 years of personal and 8 years of scientific observations I feel that I have come closer to find answers. There are many pleasurable consequences related to risk taking activities. For example, the pleasant feelings of arousal, high activity and optimal bodily and mental functioning, the praise and admiration from fellow human beings, a realization of mastery and control over my own life and for a short time even control of my death, or the ability to cope with dangerous situations and to turn them into a benefit for my own ego. Whenever it becomes boring I have to do something exciting, to compensate for the dullness day-to-day life sometimes inflicts on me. These are just a few of the reasons for me to go where no one has gone before and to keep on looking for risks I can master and that are challenging for my personal growth.

For me, taking risks is not result of some *death-drive*, but quite contrary, the drive to stay alive, to gain the illusion to be part of eternity, and to do it in the most pleasurable but controlled way possible (especially as I have just become father of the most wonderful daughter). When I look at the long and successful history of risk taking behavior from the first humans to today's humans, then I am sure that risk taking behavior will continue to be with me and all of us. Given the insatiable drive of humans for *more*, I can only hope that our drive for more risks will be accompanied by an equally strong drive for mastery and control, so that we do not have to take the ultimate risk, just to see what happens in the *afterlife*. Maybe the consequences would be beyond the optimal level of challenge.

References

Aasman, J., Mulder, G. & Mulder, L. M. (1987). Operator effort and the measurement of heart-rate variability. *Human Factors*, *29*, 161-170.
Abbott, B.B. & Badia, P. (1986). Predictable versus unpredictable shock conditions and physiological measures of stress: a reply to Arthur. *Psychological Bulletin*, *100*, 384-387.
Adams, J.G.U. (1985). *Risk and Freedom*. Cardiff: Transport Publishing Projects.
Adams, J.G.U. (1988). Risk homeostasis and the purpose of safety regulation. *Ergonomics*, *31*, 407-428.
Adams, J.G.U. (1988). Legal restraints on drunken driving. Paper presented at EUROSAFE '88. London: July 1988.
Adler, A. (1930). *Individual Psychology*. In: C. Murchinson (Ed.), *Psychologies of 1930*. Worcester, Mas: Clark University Press.
Ahmed, S. U. (1985). Risk-Taking Propensity, Locus of Control and Entrepreneurship. *Personality and Individual Differences*, *6*, 781-782.
Ajzen, I. & Madden, T.J.(1985). Prediction of Goal Directed Behaviour: Attitudes, Intentions and Perceived Behavioral Control. *Journal of Experimental Social Psychology, 22,* 453-474.
Alderson, M.R. (1981). *International Mortality Statistics*. New York: Facts on File.
Alexander, F. (1949). The accident prone individual. *Public Health*, *64*, 357-362.
Allen, J.B., Kenrick, D.T., Linder, D.E. & Mc Call, M.A. (1989). Arousal and attraction: A response-fascilitation alternative to misattribution and negative reinforcement models. *Journal of Personality and Social Psychology*, *57*, 261-270.
Allred, K.D. & Smith, T.W. (1989). The hardy personality: Cognitive and physiological responses to evaluative threat. *Journal of Personality and Social Psychology*, *56*, 257-266.
Allport, G.W. (1937). *Personality: A Psychological Interpretation.* New York: Holt.
Amelang, M. & Breit, C. (1983). Extraversion and rapid tapping: Reactive inhibition or general cortical activation as determinants of performance differences. *Personality and Individual Differences*, *4*, 103-105.
Anderson, K.J. (1990). Arousal and the inverted-U hypothesis: A critique of Neiss's "Reconceptualizing Arousal". *Psychological Bulletin*, *107*, 96-100.

Anderson, G. & Brown, R.I.F. (1984). Real and laboratory gambling, sensation seeking and arousal. *British Journal of Psychology*, *75*, 401-410.

Andresen, B. (1986). Reizsuche- und Erlebnismotive 1: Eine psychometrische Reanalyse der SSS V Zuckerman's im Kontext der MISAP - Entwicklung. *Zeitschrift für Differentielle und Diagnostische Psychologie*, *7*, 177-203.

Andresen, B. (1990). Reizsuche- und Erlebnismotive 2: Sekundärfaktoranalysen, Invarianzprüfung und Ableitung eines MISAP-Strukturmodells. *Zeitschrift für Differentielle und Diagnostische Psychologie*, *11*, 65-92.

Apter, M.J. (1976). Some data inconsistent with the optimal arousal theory of motivation. *Perceptual and Motor Skills*, *43*, 1209-1210.

Apter, M.J. (1982). *The Experience of Motivation*. London: Academic Press.

Apter, M.J. (1984). Reversal theory and personality: A review. *Journal of Research in Personality*, *18*, 265-288.

Apter, M.J. & Smith, K.C.P. (1979). Sexual Behavior and the Theory of Psychological Reversals. In: M. Cook & G. Wilson (Eds.). *Love and Attraction - An international conference*. Oxford: Pergamon. Cited in: Apter, M.J. (1982). *The Experience of Motivation*. London: Academic Press.

Arthur, A. Z. (1986). Stress of predictable and unpredictable shock. *Psychological Bulletin*, *100*, 379-383.

Arthur, Jr., W., Barrett, G.V. & Doverspike, D. (1990). Validation of an information processing based test battery for the prediction of handling accidents among petroleum product transport drivers. *Journal of Applied Psychology*, *75*, 621-628.

Arvey, R.D., Bouchard, Jr., T.J., Segal, N.L. & Abraham, L.M. (1989). Job satisfaction: Environmental and genetic components. *Journal of Applied Psychology*, *74*, 187-192.

Aschenbrenner, K.M. (1978). Single-peaked risk preferences and their dependability on the gamble's presentation mode. *Journal of Experimental Psychology: Human Perception and Performance*, *4*, 513-520.

Aschenbrenner, K.M. (1978b). Expected risk and single-peaked preferences. *Acta Psychologica*, *42*, 343-356.

Aschenbrenner, K.M. (1981). Efficient sets, decision heuristics, and single-peaked preferences. *Journal of Mathematical Psychology*, *23*, 227-256.

Aschenbrenner, K.M. & Wurm, G.W. (1985). *Risk determines behavior: A psychometric study of risk perception and risk acceptance in scuba diving.* Unpublished Report. University of Heidelberg, W. Germany.

Aschenbrenner, K.M., Biehl, B. & Wurm, G.W. (1988). *Mehr Verkehrs sicherheit durch bessere Technik? Felduntersuchungen zur Risiko-kompensation am Beispiel des Antiblockiersystems (ABS).* Abschlußbericht an die Bundesanstalt für Straßenwesen zum Forschungsprojekt 8323: Einfluß der Risikokompensation auf die Wirkung von Sicherheitsmaßnahmen. Mannheim, W. Germany, April.

Aschenbrenner, K.M. & Biehl, B. (1993 in press) Improved safety through improved technical measures? Empirical studies regarding risk compensation processes in relation to anti-lock braking systems. In: R.M. Trimpop & G.J.S. Wilde (Eds.). *Challenges to Accident Prevention: The issue of risk compensation behaviour.* Groningen, The Netherlands: STYX Publications.

Aspinwall, L.G. & Taylor, S.E. (1992). Modeling cognitive adaptation: a longitudinal investigation of the impact of individual differences and coping on college adjustment and performance. *Journal of Personality and Social Psychology, 63,* 989-1003.

Atkinson J.W. (1957). Motivational determinants of risk taking behavior. *Psychological Review, 64,* 359-372.

Atkinson, J.W. (1982). *Old and new conceptions of how expected consequences influence actions.* In: N.T Feather (Ed.). *Expectations and Actions* (pp. 17-52). Hillsdale: Lawrence Erlbaum.

Atkinson, J.W., Bastian, J.R., Earl, R.W. & Litwin, G.H. (1960). The achievement motive, goal-setting, and probability preferences. *Journal of Abnormal Social Psychology, 60,* 27-36.

Atkinson, J.W. & Birch, D. (1978). *An Introduction to Motivation* (2nd ed.). New York: Van Nostrand Reinhold.

Aufsattler W. & Biehl, B. (1986). Auswertung des Verkehrszentralregisters zur Unfall-und Deliktbelastung verschiedener Personengruppen und seine Brauchbarkeit als Mittel der Effizienzkontrolle. Abschlußbericht zum Forschungsprojekt FP 7317/2.

Averill, J.R. (1973). Personal control over aversive stimuli and its relationship to stress. *Psychological Bulletin, 80,* 286-303.

Baddley, A. D. (1972). Selective attention and performance in dangerous environments. *British Journal of Psychology, 63,* 537-546.

Bain, A. (1875) *The Emotions and the Will* (3rd Ed.). London: Longmans Green.

Bainbridge, L. (1978). Forgotten alternatives of skill and workload. *Ergonomics, 21*, 109-114.

Bajgier, S.M. & Moskowitz, H. (1982). An interactive model of attitude and risk/benefit formation regarding social issues. *Policy Sciences, 14*, 257-278.

Baker, L.A. & Daniels, D. (1990). Nonshared environmental influences and personality differences in adult twins. *Journal of Personality and Social Psychology, 58*, 103-110.

Baldwin, J.D. (1984). A balanced emphasis on environmental influences. *Behavior and Brain Science, 7*, 434-435.

Ball, I.L., Farnill, D. & Wangeman, J.F. (1984). Sex and age differences in sensation seeking: Some national comparisons. *British Journal of Psychology, 75*, 257-265.

Ball, S.A., & Zuckerman, M. (1992). Sensation Seeking and Selective Attention: Focused and Devided Attention on a Dichotic Listening Task. *Journal of Personality and Social Psychology, 63*, 825-831.

Bandura, A. (1977). Self-efficacy: Toward a unifying theory of behavioral change. *Psychological Review, 84*, 191-215.

Bandura, A. & Wood, R. (1989). Effect of perceived controllability and performance standards on self-regulation of complex decision making. *Journal of Personality and Social Psychology, 56*, 805-814.

Bandura, A., Ross, D & Ross, S. (1963). A comparative test of status envy, social power, and secondary reinforcement theories of identification learning. *Journal of Abnormal and Social Psychology, 67*, 527-534.

Barratt, E.S. (1984). Personality Traits: Causation, Correlation, or Neo-Bayesian. *Behavior and Brain Science, 7*, 435-436.

Barratt, E.S. & Patton, J.H. (1983). Impulsivity: Cognitive, Behavioral, and Psychophysiological Correlates. In: M. Zuckerman (Ed.). *Biological Bases of Sensation Seeking, Impulsivity, and Anxiety* (pp. 72-116). Hillsdale: Lawrence Erlbaum.

Battmann, W. (1984). Der jugendliche Motorradfahrer zwischen Hedonismus und Ökologiebewußtsein: Wenn das Gute nicht nützlich ist. In: A. Striksrud (Ed.). *Jugend und Werte.* Weinheim: Beltz.

Baum, A. & Fleming, I. (1993). Implications of psychological research on stress and technological accidents. *American Psychologist, 48*, 665-672.

Baum, A. Gruneberg, N.E. & Singer, J. (1982).The use of psychological and neuroendocrinological measures in the study of stress. *Health Psychology, 1*, 216-236.

Beach, B.H. & Beach, L.R. (1982). *Expectancy-based decision schemes: sidesteps towards applications*. In: N.T. Feather (Ed.) *Expectations and Actions* (pp. 365-394). Hillsdale: Lawrence Erlbaum.
Bell, D.E. (1985). Disappointment on Regret. *Operations Research, 33*, 1-27.
Becker, W. (1993). Verkehrsaufklärungsmaßnahmen zur Verbesserung der Verkehrssicherheit: Wirkungsdeterminanten und Erfolgskriterien. *Zeitschrift für Verkehrssicherheit, 39,* 110-120.
Beebe-Center, J.G. (1932). *Pleasantness and Unpleasantness*. New York: Van Nostrand.
Berlyne, D.E. (1960). *Conflict, arousal and curiosity*. New York: McGraw-Hill.
Berlyne, D.E. (1971). *Aesthetics and Psychobiology*. New York: Appleton-Century-Crofts.
Bernoulli, D. (1738). Specimen theoriae novae de mensura sortis. *Commentarii academiae scientiarium imperalis Petropolitanae*, *5*, 175-192. Transl. 1967 as: *Exposition of a new theory on the measurement of risk*. Farnsborough Hants, England: GreggPress.
Best, C.L. & Kilpatrick, D.G. (1977). Psychological profiles of rape crisis counselors. *Psychological Reports*, *40*, 1127-1134.
Birenbaum, M. & Montag, I. (1987). On the replicability of the factorial structure of the sensation seeking scale. *Personality and Individual Differences*, *8*, 403-408.
Bishop, D.V.M. (1977). The P scale and psychosis. *Journal of Abnormal Psychology*, *86*, 127-134.
Björkman, M. (1984). Decision making, risk taking and psychological time: Review of empirical findings and psychological theory. *Scandinavian Journal of Psychology*, *25*, 31-49.
Blackburn, R. (1969). Sensation seeking, impulsivity and psychopathic per sonality. *Journal of Consulting and Clinical Psychology*, *33*, 571-574
Blackburn R. (1975). Aggression and the EEG: a quantitative analysis. *Journal of Abnormal Psychology*, *84*, 358-364.
Blackburn, R. (1979). Cortical and autonomic arousal in primary and secondary psychopaths. *Psychophysiology*, *16*, 143-150.
Blanchard, D.C. & Blanchard, R.J. (1984). Affect and aggression: an animal model applied to human behavior. In: D.C. Blanchard & R.J. Blanchard (Eds.). *Advances in the Study of Aggression* (pp. 1-58). New York: Academic Press.
Blanchard, R.J. & Blanchard, D.C. (1987). Ethoexperimental approaches to the study of fear. *Psychological Records*.

Blanchard, D.C. & Blanchard, R.J. (1988). Ethoexperimental Approaches to the biology of emotion. *Annual Review of Psychology*, *39*, 43-68.
Blankenship, V. (1982). The relationship between consummatory value of success and achievement-task difficulty. *Journal of Personality and Social Psychology*, *42*, 901-914.
Blankenship, V. (1986). Substitution in Achievement Behavior. In: J. Kuhl & J.W. Atkinson (Eds.). *Motivation, Thought, and Action* (pp. 186-202). New York: Praeger Scientific.
Blankenship, V. (1987). A computer-based measure of resultant achievement motivation. *Journal of Personality and Social Psychology*, *53*, 361-372.
Blascovitch, J., Brennan, K., Tomaka, J., Kelsey, R.M., Hughes, P., Coad, M.L. & Adlin, R. (1992). Affect intensity and cardiac arousal. *Journal of Personality and Social Psychology*, *63*, 164-174.
Blitzer, C.M., Johnson, R.J., Ettlinger, C.F. & Aggeborn, K. (1984). Downhill skiing injuries in children. *American Journal of Sports Medicine*, *12*, 142-147.
Block, J. (1977). The Eysencks and psychoticism. *Journal of Abnormal Psychology*, *86*, 653-654.
Blomquist, G. (1986). A utility maximization model of driver traffic safety behavior. *Accident Analysis and Prevention*, *18*, 371-375.
Bouchard, Jr., T.J. (1984). Twins reared together and apart: what they tell us about human diversity. In: S.W. Fox (Ed.). *Individuality and Determinism Chemical and Biological Bases* (pp. 147-178). New York: Plenum Publishing Corporation.
Bouchard, Jr., T.J. (1986). Diversity, development and determination: a report on identical twins reared apart. In: M. Amelang (Ed.). *Bericht über den 35. Kongreß der Deutschen Gesellschaft für Psychologie in Heidelberg* (pp.417-433). Göttingen:Verlag für Psychologie, Hogrefe.
Boucsien, W. & Hoffmann, G. (1979). A direct comparison of the skin conductance and skin resistance methods. *Psychophysiology, 16*, 66-70.
Bouter, L.M., Knipschild, P.G., Feij, J.A. & Volovics, A. (1988). Sensation seeking and injury risk in downhill skiing. *Personality and Individual Differences*, *9*, 667-673.
Bragg, B.W.E. & Finn, P. (1982). Younger driver risk-taking research: Technical report of experimental study. Report prepared for National Highway Safety Administration, U.S. Dot, HS-806-375.

Brener, J. (1987). Behavioral energetics: some effects of uncertainity on the mobilization and distribution of energy. *Psychophysiology*, *24*, 499-512.

Brenner, R. (1983). *History-The Human Gamble*. Chicago: The University of Chicago Press.

Bristow, J.M. (1981). EPQ correlates of the personality deviance scale. *Personality and Individual Differences*, *2*, 350-352.

Brown, I.D. (1982). Exposure and experience are a confounded nuisance in research on driver behavior. *Accident Analysis and Prevention*, *14*, 345-352.

Brown, I.D. & Copeman, A.K. (1975). Drivers' attitudes to the serious ness of road traffic offenses considered in relation to the design of sanctions. *Accident Analysis and Prevention*, *7*, 15-26.

Brown, I.D. & Groeger, J.A. (1988). Risk perception and decision taking during the transition between novice and experienced driver status. *Ergonomics*, *31*, 585-597.

Brown, J.D. & Huffman, W.J. (1972). Psychophysiological measures of drivers under actual driving conditions. *Journal of Safety Research*, *4*, 172-178.

Brown, L.T., Ruder, V.G., Ruder, S.H. & Young, S.D. (1974). Stimulation seeking and the change seeker index. *Journal of Consulting and Clinical Psychology*, *42*, 311.

Bruner, J. (1992). Another Look at New Look 1. *American Psychologist*, *47*, 780-783.

Buchanan, C.M., Eccles, J.S & Becker, J.B (1992). Are adolescents the victims of raging hormons: evidence for activational effects of hormons on moods and behavior at adolescence. *Psychological Bulletin*, *111*, 62-107.

Buck, R. (1980). Nonverbal behavior and the theory of emotion: The facial feedback hypothesis. *Journal of Personality and Social Psychology*, *38*, 811-824.

Buck, R. (1985). Prime Theory: An integrated view of motivation and emotion. *Psychological Review*, *92*, 389-413.

Bullock, W.A. & Gilliland, K. (1993). Eysenck's Arousal Theory of Introversion - Extraversion: A Converging Measures Investigation. *Journal of Personality and Social Psychology*, *64*, 113-123.

Bulman, R.J. & Wortman, C.B. (1977). Attributions of blame and coping in the "real world": Severe accident victims react to their lot. *Journal of Personality and Social Psychology*, *35*, 351-363.

Burger, J.M. (1986). Desire for control: The effects of familiarity and sequence of outcomes. *Journal of Research in Personality*, *20*, 66-76.
Burger, J.M. (1987). Increased performance with increased personal control: A self-presentation interpretation. *Journal of Experimental Social Psychology*, *23*, 350-360.
Burger, J.M. (1989). Negative reactions to increases in perceived personal control. *Journal of Personality and Social Psychology*, *56*, 246-256.
Burger, J.M. & Cooper, H.M. (1979). The Desirability of Control. *Motivation and Emotion*, *3*, 381-391.
Burger, J.M. & Schnerring, D.A. (1982). The effects of desire for control and extrinsic rewards on the illusion of control and gambling. *Motivation and Emotion*, *6*, 329-335.
Burger, J.M. & Smith, N. G. (1985). Desire for control and gambling behavior among problem gamblers. *Personality and Social Psychology Bulletin*, *11*, 145-152.
Burkardt, F. (1991). Die Psychologie in der Ausbildung der Sicherheits fachkräfte. In: B. Zimolong & R. Trimpop. *Psychologie der Arbeitssicherheit* (pp. 521-534). Heidelberg: Asanger.
Burkardt, F. (1981). *Information und Motivation zur Arbeitssicherheit.* Wiesbaden: Universum.
Burkhart, B.R., Schwarz, R.M. & Green, S.B. (1978). Relationships between dimensions of anxiety and sensation seeking. *Journal of Consulting and Clinical Psychology*, *46*, 194-195.
Burns, P.C. & Wilde, G.J.S. (1993). Risk taking in male taxi drivers: Relationships among personality, observational data and driver records. Queen`s University. Kingston, Ontario, Canada.
Busemeyer, J.R. & Townsend, J.T. (1993). Decision field theory: A dynamic-cognitive approach to decision making in an uncertain environment. *Psychological Review*, *100*, 432-459.
Buss, A.H. (1989). Personality as Traits. *American Psychologist*, *44*, 1378-1388.
Buss, D.M. (1983). Evolutionary biology and personality psychology: Implications of genetic variability. *Personality and Individual Differences*, *4*, 51-63.
Buss, D.M. (1984). Evolutionary biology and personality Psychology: Toward a conception of human nature and individual differences. *American Psychologist*, *39*, 1135-1147.
Buss, D.M. (1984 b). Toward a psychology of person-environment (PE) correlation: The role of spouse selection. *Journal of Personality and Social Psychology*, *47*, 361-377.

Buss, D.M. (1987). Selection, evocation, and manipulation. *Journal of Personality and Social Psychology, 53*, 1214-1227.
Buss, D.M. (1988). The Evolution of Human Intrasexual Competition: Tactics of Mate Attraction. *Journal of Personality and Social Psychology, 54*, 616-628.
Buss, D.M. (1991). Evolutionary Personality Psychology. *Annual Review of Psychology, 42*, 459-491.
Buss, D.M. & Barnes, M. (1986). Preferences in Human Mate Selection. *Journal of Personality and Social Psychology, 50*, 559-570.
Buss, D.M. & Schmitt, D.P. (1993). An Evolutionary Perspective on Human Mating. *Journal of Personality and Social Psychology, 100*, 204-232.
Cacioppo, J.T. & Tassinary, L.G. (1990). Inferring psychological significance from physiological signals. *American Psychologist, 45*, 16-28.
Cadoret, R.J., Cain, C.A. & Crowe, A.A. (1983). Evidence for gene-environment interaction in the development of adolescent antisocial behavior. *Behavioral Genetics, 13*, 301-310.
Cameron, T.L. (1982). Drinking and Driving among American Youths: Beliefs and Behaviors. *Drug and Alcohol Dependence, 10*, 1-33.
Cameron, T.L. (1983). *Psychosocial and behavioral factors in the prevention of young driver accidents.* Paper presented at the International Symposium on Young Driver Accidents, Banff, CDN.
Campbell, J.B. (1983). Differential relationships of extraversion, impulsivity, and sociability to study habits. *Journal of Research in Personality, 17*, 308-314.
Cannon, W.B. (1927). The James-Lange theory of emotion: A critical examination and an alternative theory. *American Journal of Psychology, 39*, 106-124.
Cannon, W.B. (1929). *Bodily Changes in Pain, Hunger, Fear and Rage.* New York: Appleton.
Cannon, W.B. (1932). *The Wisdom of the Body.* New York: Norton.
Carr, A.J. (1984). *The Modifying Effects of Actual and Potential Stressor Control on Physiological and Self-Reported Stress Responses.* Unpublished Ph.D. Thesis. Queen's University, Kingston, Canada.
Carr, A.J. & Wilde, G.J.S. (1988) Effects of actual and potential stressor control on physiological and self-reported stress response. *Journal of Social and Clinical Psychology, 6*, 371-387.
Cattell, R.B. (1957). *Personality, and motivation structure and measurement.* New York: Harcourt, Brace & World.

Chawalisz, K., Diener, E. & Gallagher, D. (1988). Autonomic arousal feedback and emotional experience: Evidence from the spinal cord injured. *Journal of Personality and Social Psychology*, *54*, 820-828.
Christensen-Szalanski, J.J.J., Beck, D.E., Christensen-Szalanski, C.M., & Koepsell, T.D. (1983). Effects of expertise and experience on risk judgments. *Journal of Applied Psychology*, *68*, 278-284.
Clark, A. (1984). The logic of the comparative approach. *Behavior and Brain Science*, *7*, 437-438.
Clark, W.C. (1980). Witches, floods, and wonder drugs: Historical perspectives on risk management. In: R.C. Schwing & W.A. Jr. Albers. (Eds.). *Societal Risk Assessment: How safe is safe enough?* New York: Plenum Press.
Clarke, K.S. (1966). Calculated risk of sport fatalities. *Journal of the American Medical Association*, *197*, 172-175.
Clavier, R.M. (1984). Monoamines an human traits: A nice idea, but... . *Behavior and Brain Science*, *7*, 438-439.
Clement, R. & Jonah, B.A. (1984). Field dependence, sensation seeking and driving behavior. *Personality and Individual Differences*, *5*, 87-93.
Clifford, M.M. (1991). Risk Taking: Theoretical, Empirical, and Educational Considerations. *Educational Psychologist*, *26*, 263-297.
Cohen, J. (1960). *Chance, Skill, and Luck*. Middlesex: Penguin Books.
Cohen, L.H. (1982). Life change and the sensation seeking motive. *Personality and Individual Differences*, *3*, 221-222.
Cole, G.A. & Stephen, B.W. (1982). The Risk of Aggregation. Risk Analysis, 2, 243-247.
Colbourn, C.G. (1978). Perceived risk as a determinant of driver behavior. *Accident Analysis and Prevention*, *10*, 131-141.
Conybeare, J.A.C. (1980). Evaluation of automobile safety regulations: The case of compulsory seat-belt legislation in Australia. *Policy Sciences*, *12*, 27-39.
Coombs, C. H. (1972). *A Review of Mathematical Psychology of Risk*. Michigan Mathematical Psychology Program, MMPP-72-6. University of Michigan: Ann Arbor.
Coombs, C. H. (1975). Portfolio theory and the measurement of risk. In: M. I. Kaplan & S. Schwartz (Eds.). *Human Judgement and Decision Processes*. New York: Academic Press.
Coombs, C. H. & Huang, L. C. (1970). Test of a portfolio theory of risk preference. *Journal of Experimental Psychology*, *85*, 23-39.

Cooper, W.S. (1987). Decision theory as a branch of evolutionary theory: A biological derivation of the Savage axioms. *Psychological Review, 94*, 395-411.
Cooper, W.H. & Richardson, A.J. (1986). Unfair Comparisons. *Journal of Applied Psychology, 71*, 179-184.
Corcoran, D. W. J. (1965). Personality and the inverted - U-relation. *British Journal of Psychology, 56*, 267 - 273.
Covello, V.T., Menkes, J. & Nehnevasja, J. (1982). Risk analysis, philosophy and the social and behavioral sciences: Reflections on the scope of risk analysis research. *Risk Analysis, 2*, 53-58.
Covello, V.T. & Mumpower,J. (1985). Risk analysis and risk management: An historical perspective. *Risk Analysis, 5*, 103-120.
Cowles, M. & Davis, C. (1985). Strength of the nervous system and reversal theory. In: M.J. Apter, D. Fontana & S. Murgatroyd (Eds.). *Reversal Theory: Applications and Developments*. Cardiff, Wales: University College Press.
Cownie, A.R. & Calderwood, J.H. (1966). Feedback and accident control. *Operational Research Quarterly, 17*, 253-262.
Cox, T. & Mackay, C. (1985). The measurement of self-reported stress and arousal. *British Journal of Psychology, 76*, 183-186.
Cox, S. & Cox, T. (1991). The structure of employee attitudes to safety: a European example. *Work and Stress, 5*, 93-106.
Craig, R. & Siegel, P. (1979). Does negative affect beget positive affect? A test of opponent process theory. *Bulletin of the Psychonomic Society, 14*, 404-406.
Crawford, C.B. (1989). The theory of evolution: Of what value to psychology? *Comparative Psychology, 103*, 4-22.
Crawford, C.B. & Anderson, J.L. (1989). Sociobiology: An environmentalist discipline? *American Psychologist, 44*, 1449-1459.
Csikszentmihalyi, M: (1985). *Das Flow Erlebnis*. Stuttgart: Klett.
Csikszentmihalyi, M: (1988) The flow experience and its significance for human psychology. In: M.Csikszentmihalyi & I.S. Csikszentmihalyi (Eds.). *Optimal Experience*. New York: Cambridge University press.
Cunningham, M.R. (1988). What do you do when you're happy or blue? Mood, expectancies, and behavioral interest. *Motivation and Emotion, 12*, 309-331.
Dabbs Jr., J.M., De La Rue, D. & Williams, P. (1990). Testosterone and occupational choice: Actors, miners and other men. *Journal of Personality and Social Psychology, 59*, 1261-1265.

Daimler-Benz A.G. (1986) Tendenzen bei Kraftfahrzeugunfällen. Data presented at a lecture in the Bergische Universität Wuppertal.
Dake, K. (1991). Orienting dispositions in the perception of risk: an analysis of contemporary worldviews and cultural biases. *Journal of Cross-Cultural Psychology, 22*, 61-82.
Daly, M. & Wilson, M.I. (1982). Homicide and kinship. *American Anthropologist, 84*, 372-378.
Daly, M. & Wilson, M.I. (1988). *Homicide*. Hawthorne: Aldine de Gruyter.
Darwin, C. (1859/1968) *The Origin of Species*. London: Penguin Books.
Darwin, C. (1871). *The Descent of Man and selection in relation to sex*. London: Murray.
Darwin, C. (1872/1934). *Expressions of Emotions in Man and Animal*. London: Watts.
Davis, G.A., Peterson, J.M. & Farley, F.H. (1973). Attitudes, motivation, sensation seeking and belief in ESP as predictors of real creative behavior. *Journal of Creative Behavior, 8*, 31-39.
Dawkins, R. (1976). *The Selfish Gene*. New York, Oxford: Oxford University Press.
Dawkins, R. (1986). *The Blind Watchmaker*. Essex: Longman Scientific & Technical.
DeCharms, R. (1968). *Personal Causation: The internal affective determinants of behavior*. New York: Academic Press.
Deci, E.L. (1975). *Intrinsic Motivation*. New York: Plenum Press.
Deci, E.L. & Ryan, R. M. (1985). The general causality orientations scale: self-determination in personality. *Journal of Research in Personality, 19*, 109-134.
Deci, E.L. & Ryan, R.M. (1987). The Support of Autonomy and the Control of Behavior. *Journal of Personality and Social Psychology, 53*, 1024-1037.
DeHaven, K.E. & Lintner, D.M. (1986). Athletic injuries: Comparison by age, sport, and gender. *American Journal of Sports Medicine, 14*, 218-224.
Deldin, P.J. & Levin, I.P. (1986). The effect of mood induction in a risky decision-making task. *Bulletin of the Psychonomic Society, 24*, 4-6.
Department of Transport, United Kingdom (1990). *Road Accidents Great Britain, 1989*. London: H.M.S.O.
Derryberry, D. (1988). Emotional influences on evaluative judgments: Roles of arousal, attention, and spreading activation. *Motivation and Emotion, 12*, 23-55.

Dewsberry, D.A. (1991). Psychobiology. *American Psychologist*, 46, 198-204.
Dickenson, R. (1988). *A Study of Risk Homeostasis*. M.A. Thesis, Queen's University, Kingston, Canada.
Diekmeyer, V. (1968). *Kind und Unfall: Verkehrssoziographie für München 1966*. München: FRG.
Diener, C.I. & Dweck, C.S. (1978). An analysis of learned helplessness: Continuous changes in performance, strategy and achievement cognitions following failure. *Journal of Personality and Social Psychology*, *36*, 451-462.
Diener, C.I. & Dweck, C.S. (1980). An analysis of learned helplessness: II. The processing of success. *Journal of Personality and Social Psychology*, *36*, 451-462.
Dienstbier, R.A. (1989). Arousal and physiological toughness: Implications for mental and physical health. *Psychological review*, *96*, 84-100.
Dilalla, L.F. & Gottesman, I.I. (1991) Biological and genetic contributors to violence - Widoms untold tale. *Psychological Bulletin*, *109*,125-129.
Dörner, D. (1989). *Die Logik des Mißlingens*. Stuttgart: Rowohlt.
Doll, J. & Ajzen, I. (1992). Accessibility and stability of predictors in the theory of planned behavior. *Journal of Personality and Social Psychology*, *63*, 754-765.
Douglas, M. & Wildavsky, A. (1982). *Risk and Culture*. Berkeley: University of California Press.
Dragutinovich, S. (1987). Stimulus intensity reducers: are they sensation seekers, extraverts, and strong nervous types? *Personality and Individual Differences*, *8*, 693-704.
Drewsberry, D.A. (1991). Psychobiology. *American Psychologist*, *46*, 198-205.
Duane, D. (1988). *Spock's World*. Paramount Pictures Corporation.
Duffy, E. (1932). The relationship between muscle tension and quality of performance. *American Journal of Psychology*, *44*, 535-546.
Duffy, E. (1957). The psychological significance of the concept of arousal or activation. *Psychological Review*, *64*, 265-275.
Duffy, E. (1962). *Activation and Behavior*. New York: Wiley.
Dunbar, F. (1947). *Mind and Body Psychosomatic Medicine*. New York: Random House.
Dvorak, H. & Linke, J. (1976). Häufigkeit und Fahrlässigkeit von Unfällen als Kriterien zur Testvalidierung. *Zeitschrift für Verkehrssicherheit*, *22*, 59-63.

Dweck, C.S. & Leggett, E.L. (1988). A social-cognitive approach to motivation and personality. *Psychological Review, 95*, 256-273.

Easterbrook, J.A. (1959). The effect of emotion on cue utilization and the organization of behavior. *Psychological Review, 66*, 187-201.

Edwards, W. (1954). The theory of decision making. *Psychological Bulletin, 51*, 380-417.

Edwards, W. (1961). Behavioral decision theory. *Annual Review of Psychology, 12*, 473-498.

Eisenberg, N., Fabes, R.A., Schaller, M., Miller, P., Carlo, G., Poulin, R., Shea, C. & Shell, R. (1991). Personality and Socialization Correlates of Vicarious Emotional Responding, *Journal of Personality and Social Psychology, 61*, 459-470.

Ekman, P., Sorenson, E.R. & Friesen, W.V. (1969). Pan-cultural elements in facial displays of emotions. *Science, 164*, 86-88.

Ekman, P., Friesen, W.V. & Simons, R.C. (1985). Is the startle reaction an emotion? *Journal of Personality and Social Psychology, 49*, 1416-1426.

Elander, J., West, R. & French, D. (1993). Behavioral correlates of individual differences in road-traffic crash risk: an examination of methods and findings. *Psychological Bulletin, 113*, 279-294.

Elliott, E.S. & Dweck, C.S. (1988). Goals: An approach to motivation and achievement. *Journal of Personality and Social Psychology, 54*, 5-12.

Embrey, D.E. (1987). SLIM-MAUD: The assessment of human error pro babilities using an interactive computer based approach. In: J. Haw good & P. Humphreys (Eds.). *Effective Decision Support Systems* (pp. 20-32). Aldershot, UK: Technical Press.

Emmons, R.A. (1992). Abstract versus concrete goals: personal striving level, physical illness, and psychological well-being. *Journal of Personality and Social Psychology, 62*, 292-300.

Emmons, R.A. & Diener, E. (1986). Situation selection as a moderator of response consistency and stability. *Journal of Personality and Social Psychology, 51*, 1013-1019.

Emmons, R.A. & King, L.A. (1988). Conflict among personal strivings: Immediate and long-term implications for psychological and physical well-being. *Journal of Personality and Social Psychology, 54*, 1040-1048.

Emmons, R.A. & King, L.A. (1989). Personal striving differentiation and affective reactivity. *Journal of Personality and Social Psychology, 56*, 478-484.

Emmons, T.D. & Webb, W.W. (1974). Subjective correlates of emotional responsivity and stimulation seeking in psychopaths, normals, and acting-out neurotics. *Journal of Consulting and Clinical Psychology, 42*, 620.

English, J. & Allison, J. (1993). Chance and Rationality. *Psychological Review, 100*, 535-545.

Enzle, M.E. & Anderson, S.A. (1993). Surveillant intentions and intrinsic motivation. *Journal of Personality and Social Psychology, 64*, 257-266.

Epstein, S. (1967). Toward a unified theory of anxiety. In: B. Maher (Ed.). *Progress in experimental personality research, (Vol.4)*. New York: Academic Press. Cited in Solomon, R. (1980). The opponent-process theory of acquired motivation: The costs of pleasure and the benefits of pain. *American Psychologist, 8*, 691-712.

Epstein, S. (1973). Expectancy and magnitude of reaction to a noxious UCS. *Psychophysiology, 10*, 100-107.

Epstein, S. & Roupenian, A. (1970). Heart rate and skin conductance during experimentally induced anxiety: The effect of uncertainty about receiving a noxious stimulus. *Journal of Personality and Social Psychology, 16*, 20-28.

Erderlyi, M.H. (1992). Psychodynamics and the Unconsciousness. *American Psychologist, 47*, 784-787.

Eron, L.D. (1984). The development of aggressive behavior from the perspective of a developing behaviorism. *American Psychologist*, 12, 135-142.

Eron, L.D., Huesman, L.R., Lefkowitz, M.M. & Walder, L.O. (1972). Does television violence cause aggression? *American Psychologist, 5*, 435-442.

Ettenson, R.T. & Coughlin, R.C (1982). Effects of type of payoff and instructions on individual risk-taking behavior. *Psychological Reports, 51*, 855-860.

Evans, G.W. & Carrere, S. (1991). Traffic congestion, perceived control, and psychophysiological stress among urban bus drivers. *Journal of Applied Psychology, 76*. 658-663.

Evans, K.M., Kienast, P. & Mitchell, T.R. (1988). The effects of lottery incentive programs on performance. *Journal of Organizational Behavior Management, 9*, 113-117.

Evans, L. & Wasielewski, P. (1982). Do accident-involved drivers exhibit riskier everyday driving behavior. *Accident Analysis and Prevention, 14*, 57-64.

Evans, L. & Wasielewski, P. (1983). Risky driving related to vehicle characteristics. *Accident Analysis and Prevention, 15*, 121-136.
Evans, L., Wasielewski, P. & von Buseck, C.R. (1983). Compulsatory seat-belt usage and driver risk taking behavior. *Human Factors, 24*, 41-48.
Evans, L. (1986). Risk homeostasis theory and traffic accident data. *Risk Analysis, 6*, 81-94.
Evans, L. (1991). *Traffic safety and the driver*. New York: Van Nostrand Reinhold.
Eysenck, H.J. (1947). *Dimensions of Personality*. New York: Praeger.
Eysenck, H.J. (1952). *The Scientific Study of Personality*. NewYork: Praeger.
Eysenck, H.J. (1957). *The dynamics of anxiety and hysteria*. New York: Praeger.
Eysenck, H.J. (1967). *The biological basis of personality*. Springfield, Ill: Thomas.
Eysenck, H.J. (1984). The comparative approach in personality study. *Behavior and Brain Science, 7*, 440-441.
Eysenck, H.J. & Eysenck, S.B.G. (1969). *Personality Structure and Measurement*. London: Routledge & Kegan Paul.
Eysenck, H.J. & Eysenck, S.B.(1975). *Manual for the Eysenck Personality Questionnaire*. San Diego: Educational and Industrial Testing Service.
Eysenck, H.J. & Eysenck, S.B.G. (1976). *Psychoticism as a dimension of personality*. London: Hodder & Stoughton.
Eysenck, S.B.G., Eysenck, H.J. & Barrett (1985). A revised version of the psychoticism scale. *Personality and Individual Differences, 6*, 21-29.
Eysenck, H.J. & Levey,A. (1972). Conditioning, introversion-extraversion and the strength of the nervous system. In: V.D. Nebylitsyn & J.A. Gray (Eds.). *Biological Bases of Individual Behavior*. New York: Academic Press.
Eysenck, S.B.G. & Long, F.Y. (1986). A cross-cultural comparison of personality in adults and children: Singapore and England. *Journal of Personality and Social Psychology, 50*, 124-130.
Eysenck, S., Pearson, P.R., Easting, G. & Allsop, J.F. (1985). Age norms for impulsiveness, venturesomeness and empathy in adults. *Personality and Individual Differences, 6*, 613-619.
Eysenck, S. & Zuckerman, M. (1978). The relationship between sensation-seeking and Eysenck's dimensions of personality. *British Journal of Psychology, 69*, 483-487.

Farley, F.H. (1986). The big T in personality. *Psychology Today*, *20*, 44-52

Farley, F.H. & Farley, S.V. (1972). Stimulus-seeking motivation and delinquent behavior in institutionalized delinquent girls. *Journal of Consulting and Clinical Psychology*, *39*, 94-97.

Feather, N.T. (Ed.) (1982). Actions in relation to expected consequences: An overview of a research program. In: N.T. Feather (Ed.). *Expectations and Actions* (pp. 53-96). Hillsdale: Lawrence Erlbaum.

Feingold, A. (1993). Gender Differences in Mate Selection Preferences: a Test of the Parental Investment Model. *Psychological Bulletin*, *112*, 125-139.

Fenz, W. (1975). Strategies for coping with stress. In: I. Sarason & C. Spielberger (Eds.). *Stress and Anxiety*. New York: Wiley.

Fernandez-Seara, J.L. (1978). Psychology of the automobile driver: Personality factors of drivers with multiple accidents. *Revista de Psicologia General y Aplicanda*, *33*, 217-228.

Festinger, L. (1957). *A Theory of Cognitive Dissonance*. Evanston, Ill. Row Peterson.

Finn, P. & Bragg, B.W.E. (1986). Perception of the risk of an accident by young and older drivers. *Accident Analysis and Prevention*, *18*, 289-298.

Fishbein, D.H. (1990). Biological perspectives in criminology. *Criminology*, *28*, 27-72.

Fishburn, P. G. (1970). *Utility Theory for Decision Making*. New York: Wiley.

Fishburn, P.G. (1981). Subjective expected utility: A review of normative theories. *Theory and Decision*, *13*, 139-199.

Fishburn, P.C. (1988). Expected utility: An anniversary and a new Era. *Journal of Risk and Uncertainty*, *1*, 267-283.

Fischhoff, B. (1977). Cost Benefit Analysis and the Art of Motorcycle Maintenance. *Policy Science*, *8*, 177-202.

Fischhoff, B., Furby, L. & Gregory, R. (1987). Evaluating voluntary risks of injury. *Accident Analysis and Prevention*, *19*, 51-62.

Fischhoff, B., Goitein, B. & Shapira, Z. (1982). The experienced utility of expected utility approaches. In: Feather, N.T. (Ed.). *Expectations and Actions* (pp.315-340). Hillsdale: Lawrence Erlbaum.

Fischhoff, B., Lichtenstein, S., Slovic, P., Derby, S. & Keeney, R.L. (1981). *Acceptable risk*. Cambridge: Cambridge University Press

Fischhoff, B., Lita, F. & Gregory, R. (1987). Evaluating voluntary risks of injury. *Accident Analysis and Prevention*, *19*, 51-62.

Fischhoff, B., Slovic, P., Lichtenstein, S., Read, S. & Combs, B. (1978). How safe is safe enough? A psychometric study of attitudes towards technological risks and benefits. *Policy Science*, *9*, 127-152.

Fischhoff, B., Watson, S.R. & Hope, C. (1984). Defining Risk. *Policy Science*, *17*, 123-139.

Fishbein, M. & Ajzen, I. (1980). *Understanding Attitudes and Predicting Social Behavior*. New Jersey: Prentice Hall.

Fiske, D.W. & Maddi, S.R. (Eds.) (1961). *Functions of Varied Experience*. Homewood: Dorsey Press.

Fliegel, S., Franke, A., Groeger, W. & Schulte, D. (Eds.) (1980). *Verhaltenstherapeutische Standardmethoden*. Stuttgart: Kohlhammer.

Folkman, S. & Lazarus, R. S. (1985). If it changes it must be a process: Study of emotion and coping during three stages of a college examination. *Journal of Personality and Social Psychology*, *48*, 150-170.

Folkman, S. & Lazarus, R.S. (1988). Coping as a mediator of emotion. *Journal of Personality and Social Psychology*, *54*, 466-475.

Fortenberry, J.C. & Smith, L.A. (1981). A comparison of risk selections. *Human Factors*, *23*, 693-700.

Fowler, C.J., von Knorring, L. & Oreland, L. (1980). Platelet mono-amine oxidase in sensation seekers. *Psychiatric Research*, *3*, 273-276.

Fox, D.K., Hopkins, B.L. & Anger, W.K. (1987). The long-term effects of a token economy on safety performance in open-pit mining. *Journal of Applied Behavior Analysis*, *20*, 215-224.

Franken, R.E. & Rowland, G.L. (1990). Sensation seeking and fantasy. *Personality and Individual Differences*, *11*, 191-193.

Frankenhäuser, M. (1980). Coping with stress at work. *International Journal of Health Services*, *11*, 491-510.

Frankenhäuser, M. & Johansson, G. (1986). Stress at work: psychobiological and psychosocial aspects. *International Review of Applied Psychology*, *35*, 287-299.

Frankenhäuser, M. & Lundberg, U. & Forsman, L. (1980). Dissociation between sympathetic-adrenal and pituary-adrenal response to an achievement situation characterized by high-controllability: Comparison between Type A and Type B males and females. *Biological Psychology*, *10*, 79-91.

Freud, S. (1904). *Die Psychopathologie des Alltagslebens*. Transl. 1953 as: *The psychopathology of everyday life*. Standard Edition of the complete psychological works of Sigmund Freud, Vol. IV, J. Strachey (Ed.). London: Hogarth Press.

Freud, S. (1915). *Triebe und Triebschicksale.* Translated 1957 as: *Instincts and their Vicissitudes.* Standard Edition of the complete psychological works of Sigmund Freud, Vol. XIV, J.Strachey (Ed.). London: Hogarth Press.

Freud, S. (1920). *Jenseits des Lustprinzips.* Leipzig: Internationaler Psychoanalytischer Verlag. Translated 1955 as: *Beyond the Pleasure Principle.* Standard Edition of the complete psychological works of Sigmund Freud, Vol.XVIII, J.Strachey (Ed.). London: Hogarth Press.

Freud, S. & Breuer, J. (1895). *Studien über Hysterie.* Leipzig: Deuticke. Translated 1955 as *Studies on Hysteria.* Standard Edition of the complete psychological works of Sigmund Freud,Vol. II, J. Strachey (Ed.). London: Hogarth Press.

Frijda, N.H. (1986). *The Emotions.* New York:Cambridge University Press.

Frijda, N.H. (1987). Emotions, cognitive structure, and action tendency. *Cognition and Emotion, 1,* 115-144.

Frijda, N.H. (1988). The laws of emotion. *American Psychologist, 43,* 349-358.

Frijda, N.H., Kuipers, P. & ter Schure, E. (1989). Relations among emotion, appraisal, and emotional action readiness. *Journal of Personality and Social Psychology, 57,* 212-228.

Fromm, E. (1941). *Escape from Freedom.* New York: Reinhardt.

Fulker, D.W., Eysenck, S. & Zuckerman, M. (1980). A genetic and environmental analysis of sensation seeking. *Journal of Research in Personality, 14,* 261-281.

Fuller, R. (1984). A conceptualization of driving behavior as threat-avoidance. *Ergonomics, 27,* 1139-1155.

Fuller, R. (1988). On learning to make risky decisions. *Ergonomics, 31,* 519-526.

Furedy, J. J. (1987). Beyond heart rate in the cardiac psychophysiological assessment of mental effort: the T - Wave amplitude component of the Electrocardiogramm. *Human Factors, 29,* 183-194.

Furnham, A. (1984). Extraversion, sensation seeking, stimulus screening and Type A behavior pattern: The relationship between various measures of arousal. *Personality and Individual Differences, 5,* 133-140.

Furnham, A. (1987). The Structure of Economic Beliefs. *Personality and Individual Differences, 8,* 253-260.

Galen (160). *On the passions and errors of the soul.* (P.W. Harkins, translator). Columbus: Ohio State University Press, 1963.

Galton, F. (1875). The history of twins as a criterion of the relative powers of nature and nurture. *Fraser's Magazine, 92*, 566-576. Cited in: Bouchard, Jr. T.J. (1984). Twins reared together and apart: what they tell us about human diversity. In: S.W. Fox (Ed.). *Individuality and Determinism Chemical and Biological Bases* (pp.147-178). New York: Plenum Publishing Corporation.

Geller, E.S. (1984). A delayed reward strategy for large-scale motivation of safety belt use: a test of long-term impact. *Accident Analysis and Prevention, 16*, 457-436.

Geller, E.S., Paterson, L. & Talbott, E. (1982). A behavioral analysis of incentive prompts for motivating seat-belt use. *Journal of Applied Behavioral Analysis, 15*, 403-415.

Gerking, S., De Haan, M. & Schulze, W. (1988). The marginal value of job safety: A contingent valuation study. *Journal of Risk and Uncertainty, 1*, 185-199.

Gilliland, K. (1980). The interactive effect of introversion-extraversion with coffeine induced arousal on verbal performance. *Journal of Research in Personality, 14*, 482-492.

Gollwitzer, P.M. & Kinney, R.F. (1989). Effects of deliberative and implemental mind sets on illusion of control. *Journal of Personality and Social Psychology, 56*, 531-542.

Goldberg, A.I., Dar-El, E.M. & Rubin, A.E. (1991). Threat perception and the readiness to participate in safety programs. *Journal of Organizational Behavior, 12*, 109-122.

Goldman, D., Kohn, P.M. & Hunt, R.W. (1983). Sensation seeking, augmenting-reducing, and absolute auditory threshold: a strength-of-the-nervous-system perspective. *Journal of Personality and Social Psychology, 45*, 405-411.

Goodman, P.S. & Garber, S. (1988). Absenteeism and accidents in a dangerous environment: empirical analysis of underground coal mines. *Journal of Applied Psychology, 73,* 81-86.

Gough, H.G. (1957). *Manual for the California Personality Inventory*. Palo Alto, Ca.: Consulting Psychologists Press.

Grant, B.A. (1985). *Empirical Investigation of the Theory of Risk Homeostasis and the Study of Risk Perceived by Automobile Drivers.* Doctoral Thesis, Queen's University. Kingston, Ontario, Canada.

Gray, J.A. (1964). *Pavlov's Typology*. Oxford: Pergamon Press.

Gray, J.A. (1972). The psychophysiological nature of extraversion-introversion: a modification of Eysenck's theory. In: V.D. Nebylitsyn & J.A. Gray (Eds.).*Biological Bases of Individual Behavior*. New York: Academic Press.

Gray, J.A., Owen, S. Davis, N. & Tsaltas, E. (1983). Psychological and Physiological Relations between Anxiety and Impulsivity. In: Zuckerman, M. (Ed.). *Biological Bases of Sensation Seeking, Impulsivity, and Anxiety* (pp. 189-218). Hillsdale: Lawrence Erlbaum.

Greenberg, J. Pyszczynski, T., Solomon, S., Rosenblatt, A., Veeder, M., Kirkland, S. & Lyon, D. (1990). Evidence for terror management theory II: The effects of mortality salience on reactions to those who threaten or bolster the cultural worldview. *Journal of Personality and Social Psychology, 58*, 308-318.

Greenwald, A.G. (1992). Unconscious Cognition Reclaimed. *American Psychologist, 47*, 766-779.

Greenwood, M. (1950). Accident proneness. *Biometrika, 37*, 24-29.

Greenwood, M. & Woods, H.M. (1919). The incidence of industrial accidents upon individuals with special reference to multiple accidents. *Industrial Fatigue Research Board, Medical Research Committee, 4*. London: Her Brittanic Majesty's Stationary Office. Cited in: Greenwood, M. (1950). Accident proneness. *Biometrika, 37*, 24-29.

Grier, B. (1980). *One thousand years of mathematical psychology*. Paper presented at Society for Mathematical Psychology Convention, Madison, Wisconsin.

Grier, B. (1981). *The early history of the theory and management of risk*. Paper presented at the judgement and decision making group meeting, Philadelphia, Pennsylvania.

Grossberg, S. (1980). How does a brain build a cognitive code? *Psychological Review, 87*, 1-51.

Grossberg, S. (1982). Processing of expected and unexpected events: A psychophysiological theory. *Psychological Review, 89*, 529-572.

Grossberg, S. & Gutowski, W.E. (1987). Neural dynamics of decision making under risk: Affective balance and cognitive-emotional interactions. *Psychological Review, 94*, 300-318.

Gstalter, H. (1985) Workload and stress in driving. *Ergonomics International*, 8, 994-996.

Gulian, E., Matthews, G. Glendon, A.I., Davies, D.R. & Debney, L.M. (1989). Dimensions of driver stress. *Ergonomics, 32*, 585-602.

Hackenberg, U. (1985). *Ein Beitrag zur Stabilitätsuntersuchung des Systems "Fahrer-Kraftrad-Straße"*. Doctoral Dissertation, Rheinisch-Westfälische Technische Hochschule Aachen, Germany.
Haddon, W., Suchman, E.A. & Klein, D. (1964). *Accident Research.* New York: Harper & Row.
Häkkinen, S. (1979). Traffic accidents and professional driver characteristics: a follow-up study. *Accident Analysis and Prevention, 11*, 7-18.
Haight, F. A. (1986). Risk, especially risk of traffic accident. *Accident Analysis and Prevention, 18*, 359-366.
Hale, A.R. (1990). safety rules O.K:? Possibilities and limitations in be havioral safety strategies. *Journal of Occupational Accidents, 12*,3-20.
Hale, A.R. & Glendon, A.I. (1987). *Individual Behavior in the Control of Danger.* Amsterdam, Oxford, New York: Elsevier.
Halpern, B., Thompson, N., Curl, W.W., Andrews, J.R., Hunter, S.C. & Boring, J.R. (1987). High school football injuries: Identifying the risk factors. *American Journal of Sports Medicine, 15*, 316-319.
Hamilton, P. (1979). Process entropy and cognitive control: mental load in internalized thought processes. In: N. Moray (Ed.). *Mental Workload* (pp. 289-298). New York: Plenum Press.
Hamilton, P. Mulder, G. Strasser, H. & Ursin, H. (1979). Final report of physiological group. In: N. Moray (Ed.). *Mental Workload* (pp. 367-377). New York: Plenum Press.
Hamilton, V. & Warburton, D. M. (Eds.) (1979). *Human stress and cognition.* Chichester: John Wiley & Sons.
Hammer, W. (1980). *Product Safety Management and Engineering.* Englewood Cliffs, N.J.: Prentice-Hall.
Hannaman, G.W. & Spurgin, A.J. (1983). Systematic human action reliability procedure (SHARP) (Report No. NUS-4486). San Diego, CA: NUS Corporation.
Hansen, C.P. (1989). A causal model of the relationship among accidents, biodata, personality, and cognitive factors. *Journal of Applied Psychology, 74*, 81-90.
Hansson, S.O. (1989). Dimensions of Risk. *Risk Analysis, 9*, 107-112.
Harano, R.M. & Hubert, D.E. (1974). *An Evaluation of California's "good drivers" incentive program.* Research and Statistics Section, Department of Motor Vehicles, Highway Research Report B0146, Sacramento, Ca.NTS: CAL-DMV-RSS-74-46.
Hare, R.D. (1970). *Psychopathy: Theory and Research.* New York: John Wiley & Sons.

Hare, R.D. (1980). A research scale for the assessment of psychopathy in criminal populations. *Personality and Individual Differences*, *1*, 111-119.

Hare, R.D. (1982). Psychopathy and the personality dimensions of psychoticism, extraversion and neuroticism. *Personality and Individual Differences*, *3*, 35-42.

Harlow, H.F. (1953). Mice, monkeys, men and motives. *Psychological Review*, *60*, 23-32.

Harms-Ringdahl, L. (1990). On economic evaluation of systematic safety work at companies. *Journal of Occupational Accidents*, *12*, 89-98.

Harrell, W.A. (1990). Perceived Risk of Occupational Injury: Control Over Pace of Work and Blue-Collar versus White-Collar Work. *Perceptual and Motor Skills, 70*, 1351-1359.

Harter, S. (1978). Pleasure derived from challenge and the effects of receiving grades on children`s learning difficulty level choices. *Child Development, 49*, 788-799.

Hathaway, S.R. & McKinley, J.C. (1951). *The Minnesota Multiphasic Inventory Manual*. New York: Psychological Corporation.

Hay, M. (1988). *Risk Taking Strategy, Competition, and the Link between Laboratory and Real Life Risk Taking*. B.A. Thesis. Queen's University, Kingston, Ontario, Canada.

Heath, A.C. & Martin, N.G. (1990). Psychoticism as a dimension of personality: A multivariate genetic test of Eysenck and Eysenck's Psychoticism concept. *Journal of Personality and Social Psychology*, *58*,111-121.

Heath, A.C., Jardine, R., Eaves, L.J. & Martin, N.G. (1989). The genetic structure of Personality-II. Genetic item analysis of the EPQ. *Personality and Individual Differences*, *10*, 615-624.

Hebb, D.O. (1949). *The Organization of Behavior*. New York: Wiley

Hebb, D.O. (1955). Drives and the CNS (Conceptual Nervous System). *Psychological Review*, *62*, 243-254.

Heckhausen, H. (1977). Achievement motivation and its constructs: A cognitive model. *Motivation and Emotion*, *1*, 283-329.

Heckhausen, H. (1980). *Motivation und Handeln*. Heidelberg: Springer.

Heider, F. (1958). *The psychology of interpersonal relations*. New York: John Wiley & Sons.

Hendrick, I.(1943). The discussion of the "Instinct to Master". *Psycho analytic Quarterly*, *12*, 561-565.

Henslin, J.M. (1967). Crap and Magic. *American Journal of Sociology*, *73*, 316-330.

Herbart, J.F. (1816/1964). *A Textbook in Psychology*. Cambridge Mass.: Harvard University Press.
Heino, A., van der Molen, H.H. & Wilde, G.J.S. (1990). Risk-homeo static processes in car-following behaviour: Electrodermal responses and verbal risk estimates as indicators of the perceived levels of risk during a car-driving task. Groningen, NL: Traffic Research Center
Heino, A., van der Molen, H.H. & Wilde, G.J.S. (1992). Risk-homeostatic processes in car-following behaviour: Individual differences in car following and perceived risk. Groningen,NL: Traffic Research Center
Heron, W. (1957). The pathology of boredom. *Scientific American*, *196*, 52-56.
Heyman, S.R. & Rose, K.G. (1980). Psychological variables affecting SCUBA performance. In: C.A. Nadeau, W.R. Halliwell, K.M. Newell & G.C. Roberts (Eds.). *Psychology of Motor Behavior and Sport*. Champaign, Ill.: Human Kinetics Press.
Hilakivi, I., Velhati, J., Asplund, P., Sinivuo, J., Laitinen, L. & Koskenvuo, K. (1989). A sixteen-factor personality test for predicting automobile driving accidents of young drivers. *Accident Analysis and Prevention*, *21*, 413-418.
Hilgard, E.R. & Bower, G.H. (1966). *Theories of Learning*. New York: Appleton-Century-Crofts.
Hippocrates. *The Genuine Works of Hippocrates*. (F. Adams, translator). London Sydenham Society, 1849; cited from: Murray, D.J. (1988). *A History of Western Psychology* (2nd ed.), Englewood Cliffs, N.J.: Prentice-Hall.
Hiroto, D.S. & Seligman, M.E.P. (1975). Generality of learned helplessness in man. *Journal of Personality and Social Psychology*, *31*, 311-327.
Hockey, R. (1979). Stress and the cognitive components of skilled performance. In: V. Hamilton & D.M. Warburton (Eds.). *Human Stress and Cognition* (pp.141-178). Chichester: John Wiley & Sons.
Hodgdon, J.D., Bragg, B.W.E. & Finn, P. (1982). *Young driver risk-taking research: The state of the art*. Report prepared for National Highway Safety Administration, U.S. Dot, HS-805-967.
Hoffman, H.S. & Solomon, R.L. (1974). An opponent-process theory of motivation: III. Some affective dynamics in imprinting. *Learning and Motivation*, *5*, 149-164.
Hoffman, L.W. (1991). The influence of family environment on personality: accounting for sibling differences. *Psychological Bulletin*, *110*,187-203.

Holsti, O.R. (1978). Limitations of cognitive abilities in the face of crisis. In: C.F. Smart & W.T. Stanbury (Eds.). *Studies on Crisis Management*. Toronto: Butterworth.

Horn, J.M. (1983). The Texas Adoption Projekt: Adopted children and their intellectual resemblance to their biological and adoptive parents. *Child Development*, *54*, 268-275.

Howard, R., Bailey, R. & Newman, A. (1984). A preliminary study of Hare's research scale for the assessment of psychopathy in mentally abnormal offenders. *Personality and Individual Differences*, *5*, 389-396.

Howard, R.C. (1986). Psychopathy: A psychobiological perspective. *Personality and Individual Differences*, *7*, 795-806.

Howarth, C.I. (1987). Perceived risk and behavioral feedback: strategies for reducing accidents and increasing efficiency. *Work and Stress*, *1*, 61-65.

Howarth, C.I. (1988). The relationship between objective risk, subjective risk and behavior. *Ergonomics*, *31*, 527-535.

Hoyos, C. Graf (1980). Psychologische Unfall und Sicherheitsforschung. Stuttgart: Kohlhammer.

Hoyos, C. Graf. (1986). Sicherheit durch sicheres Verhalten. *Zeitschrift für Arbeitswissenschaft*, *40*, 201-207.

Hoyos, C. Graf. (1988). Mental load and risk in traffic behavior. *Ergonomics*, *31*, 571-584.

Hoyos, C. Graf & Zimolong, B. (1988). *Occupational safety and accident prevention. Behavioral strategies and methods*. Amsterdam: Elsevier Science Publisher.

Hudgens, G.A. & Fatkin, L.T. (1985). Sex differences in risk-taking: repeated sessions on a computer simulated task. *The Journal of Psychology*, *119*, 197-206.

Huguenin, R.D. (1993). Marketing in der Verkehrssicherheitsarbeit. *Zeitschrift für Verkehrssicherheit, 39*, 84-86.

Hughs, J. *Ecology in Ancient Civilizations*. University of Chicago Press. Cited in: Covello, V.T. & Mumpower, J. (1985). Risk analysis and risk management: A historical perspective. *Risk Analysis*, *5*, 103-120.

Hulbert, S.F. (1957). Drivers' GSRs in traffic. *Perceptual and Motor Skills*, *7*, 305-315.

Hull, C.L. (1943). *Principles of behavior: An introduction to behavior theory*. New York: Appleton-Century-Crofts.

Humphreys, P.C. & Wisudha, A. (1983). MAUD 4, Technical Report 83-5. Decision Analysis Unit. London: London School of Economics and Political Science.
Humphreys, M.S. & Revelle, W. (1984). Personality, motivation, and performance: a theory of the relationship between individual differences and information processing. *Psychological Review, 91*, 153-184.
Hunt, E. (1986). *The Effects of Coffeine on Risk Taking Behavior.* B.A. Thesis. Queen's University, Kingston, Ontario, Canada.
Hurt, Jr., H.H., Oullet, J.V. & Thom, D.R. (1981). Motorcycle accident cause factors and identification of countermeasures. U.S. Department of Transportation. Report no. *DOT-HS-805862.*
Hymbaugh, K. & Garrett, J. (1974). Sensation seeking among skydivers. *Perceptual and Motor Skills, 38*, 118.
Isen, A.M., Daubman, K.A. & Nowicki, G.P. (1987). Positive affect facilitates creative problem solving. *Journal of Personality and Social Psychology, 52*, 1122-1131.
Isen, A.M., Nygren, T.E. & Ashby, F.G. (1988). Influence of positive affect on the subjective utility of gains and losses: It is just not worth the risk. *Journal of Personality and Social Psychology, 55*, 710-717.
Isen, A.M. & Patrick, R. (1983). The effect of positive feelings on risk-taking: When the chips are down. *Organizational Behavior and Human Performance, 31*, 194-202.
Isen, A.M. (1990). The Influence of Positive and Negative Affect on cognitive Organization: Some Implications for development. In: N. Stein, B. Leventhal, T. Trabasso (pp. 75-94). *Psychological and Biological Processes in the Development of Emotion.* Hillsdale, NJ: Erlbaum.
Isen, A. M., Baron, R.A. (1991). Positive Affect as a Factor in Organizational Behavior. In: L.L. Cummings & B.L. Staw (pp. 1-35). *Research in Organizational Behavior.* Greenwich, CT: JAI Press.
Izard, C.E. (1977). *Human Emotions.* New York: Plenum Press.
Izard, C.E., Kagan, J., Zajonc. R.B. (Eds.) (1984). *Emotions, Cognitions, and Behavior.* Cambridge University Press.
Izard, C.E. (1984). Emotion variables as personality traits. *Behavior and Brain Science, 7*, 442-443.
Izard, C.E. (1993).Four Systems for Emotion Activation: Cognitive and Noncognitive Processes. *Psychological Review, 100*, 68-90.
Jackson, D.N. (1967). *Personality Research Form Manual.* Goshen, N.Y.: Research Psychologists Press.

Jackson, D.N. Hourany, L. & Vidmar, N.J. (1972). A four-dimensional interpretation of risk taking. *Journal of Personality*, *40*, 483-501.
Jacobi, L.L., Lindsay, D.S. & Toth, J.P. (1992). Unconscious Influences Revealed: Attention, Awareness and Control. *American Psychologist*, *47*, 802-809
Janis, I.L. & Mann, L. *Decision making: A psychological analysis of conflict, choice and commitment*. New York: Free Press.
Jannssen, W. (1988). *Risk compensation and the effect of an incenive: A laboratory study*. TNO Reportnr. IZF 1988 C-26. Leidschendam, Nl.
Jannssen, W. & Tenkink, E. (1988). Considerations on speed selection and risk homeostasis in driving. *Accident Analysis and Prevention*, *20*, 137-142.
Jannssen, W. & Tenkink, E. (1988b). Risk homeostasis theory and its critics: time for an agreement. *Ergonomics*, *31*, 429-433.
Janssens, L., Grotenhuis, H., Michiels, H. & Verhaegen, P. (1989). Social organizational determinants of safety in nuclear power plants: Operator training in the management of unforeseen events. *Journal of Occupational Accidents*, *11*, 121-129.
Janoff-Bulman & Brickman, P. (1982). *Expectations and what people learn from failure*. In: N.T. Feather (Ed.) (p.207-240). *Expectations and Actions*. Hillsdale: Lawrence Erlbaum.
Jeffery, R.W. (1989). Risk Behaviors and Health: Contrasting Individual and Population Perspectives. *American Psychologist*, *44*, 1194-1202.
Jenkins, C.D., Rosenman, R.H. & Friedman, M. (1967). Development of an objective psychological test for the determination of the coronary prone behavior pattern in employed men. *Journal of Chronic Diseases*, *20*, 371-379.
Jenkins, C.D. (1971). Psychologic and social precursors of coronary heart disease. *New England Journal of Medicine*, *284*, 307-317.
Jermier, J.M., Gaines, J. & McIntosh, N.J. (1989). Reactions to physically dangerous work: A conceptual and empirical analysis. *Journal of Organizational Behavior*, *10*, 15-33.
Jex, H.R. & Clement, W.F. (1979). Workload in manual control tasks. In: N. Moray, (Ed.) (pp.125-178). *Mental Workload*. New York: Plenum Press.
Johansson, G. & Frankenhäuser, J. (1973). Temporal factors in sympatho-adrenomedullary activity following acute behavioral activation. *Biological Psychology*, *1*, 63-73.
Johansson, G. & Aronsson, G. (1984). Stress reactions in computerized administrative work. *Journal of Occupational Behavior*, *5*, 159-181.

Johansson, G., Aronsson, G. & Lindstroem, B.O. (1978). Social Psychological and Neuroendocrine Stress Reactions in Highly Mechanized Work. *Ergonomics, 21*, 583-599.

Johnson, B.B. (1991). Risk and Culture research: Some Caution. *Journal of Cross-Cultural Psychology, 22*, 141-149.

Johnson, E.J. & Tversky, A. (1983). Affect, generalization, and the perception of risk. *Journal of Personality and Social Psychology, 45*, 20-31.

Johnson, R.J., Ettlinger, C.F., Campbell, R.J. & Pope, M.H. (1980). Trends in skiing injuries: Analysis of a 6-year study (1972 to 1978). *American Journal of Sports Medicine, 8*, 106-110.

Joly, R. & Wilde, G.J.S. (1987). The effect of BAC at .08% by volume upon skill and risk-taking tendency in a response-timing task. In: P.C. Noordzij & R. Roszbach (Eds.). *Alcohol, Drugs and Traffic Safety-T86*. Amsterdam: Elsevier Science Publishers.

Jonah, B.A. (1986). Accident risk and risk-taking behavior among young drivers. *Accident Analysis and Prevention, 18*, 255-271.

Jonah, B.A., Dawson, N.E. & Bragg, B.W.E. (1981). Predicting accident involvement with the Motorcycle Operator Risk Test. *Accident Analysis and Prevention, 13*, 307-318.

Jonah, B.A. & Dawson, N.E. (1982). The national vehicle occupant restraint survey: Attitudes towards the use of restraints by Canadians. TP3593E. Road Safety Directorate, Transport Canada, March.

Jonah, B.A. & Dawson, N.E. (1983). Predicting reported seat belt use from attitudinal and normative factors. *Accident Analysis and Prevention, 14*, 305-309.

Jonah, B.A., Dawson, N.E. & Bragg, B.W.E. (1982). Are formally trained motorcyclists safer? *Accident Analysis and Prevention, 14*, 247-255.

Jonah, B.A. & Lawson, J.J. (1984). The effectiveness of the Canadian mandatory seat belt use laws. *Accident Analysis and Prevention, 16*, 433-450.

Jones, J.W. & Wuebker, L. (1985). Development and Validation of the Safety Locus of Control Scale. *Perceptual and Motor Skills, 61*, 151-161.

Josephs, R.A., Larrick, R.P., Steele, C.M. & Nisbett, R.E. (1993). Protecting the self from the negative consequences of risky decisions. *Journal of Personality and Social Psychology, 62*, 26-37.

Jung, C.G. (1954). The aims of psychotherapy. In: *The Practice of Psychotherapy*. Collected Works, Vo.16, London: Routledge and Kegan Paul.

Jungermann, H. & Slovic, P. (1987). *Die Psychologie der Kognition und Evaluation von Risiko*. Unpublished Manuscript. Technische Universität Berlin.

Kafry, D. (1982). Sensation seeking of young children. *Personality and Individual Differences, 3*, 161-166.

Kahneman, D. & Tversky, A. (1973). On the psychology of prediction. *Psychological Review, 80*, 237-251.

Kahneman, D. & Tversky, A. (1979). Prospect theory: An analysis of decision under risk. *Econometrica, 47*, 263-291.

Kahneman, D. & Tversky, A. (1984). Choices, values, and frames. *American Psychologist, 39*, 341-350.

Kagan, J. & Snidman, N. (1991). Temperamental factors in human development. *American Psychologist, 46*, 856-862.

Kaplan, L.B., Szybillo, G.J. & Jacoby, J. (1974). Components of Perceived Risk in Product Purchase: A Cross-Validation. *Journal of Applied Psychology, 59*, 287-291.

Kaspar, F.E. & Vesper, J.J. (1976). Death anxiety in a risk-taker group. *Essence, 1*, 95-97.

Kasperson, R.E. (1986). Six propositions on public participation and their relevance for risk communication. *Risk Analysis, 6*, 275-281.

Katcher, A. H., Solomon, R.L., Turner, L.H., LoLordo, V.M., Overmier, J.B. & Rescorla, R.A. (1969). Heart rate and blood pressure responses to signaled and unsignaled shock: Effects of cardiac sympathectomy. *Journal of Comparative and Physiological Psychology, 68*, 163-174.

Katz, R.J. (1984). Sensation seeking: A clarification, a caveat, and a conjecture. *Behavior and Brain Science, 7*, 443.

Katz, R. & Wykes, T. (1985). The psychological difference between temporally predictable and unpredictable stressful events: Evidence for information control theories. *Journal of Personality and Social Psychology, 48*, 781-790.

Keeney, R.L. & Raiffa, H. (1976). *Decisions with Multiple Objectives: Preferences and Value Tradeoffs*. New York: John Wiley & Sons.

Keinan, G., Meir, E. & Gome-Nemirovsky, T. (1984). Measurement of risk takers' personality. *Psychological Reports, 55*, 163-167.

Keller, L.R., Sarin, R.K. & Weber, M. (1986). Empirical investigation of some properties of the perceived riskiness of gambles. *Organizational Behavior and Human Decision Processes, 38*, 114-130.

Kelley, H.H. (1971). *Attribution in Social Interaction*. Morristown, N.J.: General Learning Press.

Kelly, G.A. (1955). *The Psychology of Personal Constructs*. New York: Norton.

Kennedy, R.S., Bittner, A.C. Harbeson, M. & Jones, M.B. (1982). Television computer games: A new look in performance testing. *Aviation, Space, and Environmental Medicine, 53*, 49-53.

Kenrick, D.T., Groth, G.E., Trost, M.R. & Sadallah E.K. (1993). Integrating evolutionary and social exchange perspectives on relationships: Effects of gender, self-appraisal, and involvement level on mate selection criteria. *Journal of Personality and Social Psychology, 64*, 951-969.

Keppel, G. (1980). *Design and Analyses*. Prentice-Hall International.

Keren, G. (1987). Facing Uncertainty in Game of Bridge: A Calibration Study. *Organizational Behavior and Human Decision Processes, 39*, 98-114.

Kerr, J.H. & Cox, T. (1990). Cognition and mood in relation to the performance of squash tasks. *Acta Psychologica, 73*, 103-114.

Kerr, J.H. & Svebak, S. (1989). Motivational aspects of preference for and participation in risk and safe sports. *Personality and Individual Differences, 10*, 797-800.

Kihlstrom, J.F., Barnhardt, T.M. & Tataryn, D. J. (1992). The Psycho logical Unconscious: Found, Lost and Regained. *American Psychologist, 47*, 788-791.

Kiburz, D., Jacobs, R., Reckling, F. & Mason, J. (1986). Bicycle accidents and injuries among adult cyclists. *American Journal of Sports Medicine, 14*, 416-419.

King, K.B. (1984). *Coping with cardiac surgery*. Unpublished doctoral dissertation. University of Rochester, N.Y.

Kish, G.B. & Donnenworth, G.V. (1969). Interests and stimulus seeking. *Journal of Counseling Psychology, 16*, 551-556.

Kish, G.B. & Donnenwrth, G.V. (1972). Sex differences in the correlates of sensation seeking. *Journal of Consulting and Clinical Psychology, 38*, 42-49.

Klebelsberg, D. (1982). *Verkehrspsychologie*. Heidelberg: Springer.

Klebelsberg, D. (1984). Aktuelle Forschungs- und Anwendungsprobleme in der Verkehrspsychologie. *Perspektiven der Forschung, 147*, Universität Innsbruck, Austria.

Klebelsberg, D. (1987). Sicherheitsverhalten und Sicherheitseinstellung. In: E. Raab & G. Schulter (Eds.). *Perspektiven Psychologischer Forschung* (pp. 191-201). Wien: Franz Deuticke.

Klebelsberg, D. (1993). Is risk compensation a personality trait, an attitude or a behavior?. In: R.M. Trimpop & G.J.S. Wilde (Eds.). *Challenges to Accident Prevention: The issue of risk compensation behaviour.* Groningen, The Netherlands: STYX Publications.

Klebelsberg, D., Pilsz, W., & Schibalski, F. (1979). *Analyse des Fahrverhaltens in Kurven.* Forschungsbericht 7518/2 der Bundesanstalt für Straßenwesen in Köln.

Kleinhesselink, R.R. & Rosa, E.A. (1991). Cognitive representation of risk perception. *Journal of Cross-Cultural Psychology, 22*, 11-28.

Klonowicz, T., Zawadzka, G. & Zawadzki, B. (1987). Reactivity, arousal, and coping with stress. *Personality and Individual Differences, 6*, 793-798.

Knight, F.H. (1921). *Risk, Uncertainty, and Profit.* Boston, New York: Houghton Mifflin.

Kobasa, S.C. (1979). Stressful life-events, personality and health: An inquiry into hardiness. *Journal of Personality and Social Psychology, 37*, 1-11.

Kobasa, S.C. (1982). Commitment and coping in stress resistance among lawyers. *Journal of Personality and Social Psychology, 42*, 707-717.

Kobasa, S.C., Maddi, S.R. & Kahn, S. (1982). Hardiness and health: A prospective study. *Journal of Personality and Social Psychology, 42*, 168-177.

Kogan, N. & Wallach, M. A. (1964). *Risk Taking: A Study in Cognition and Personality.* New York: Holt, Rinehart, and Winston.

Kohn, P.M. & Coulas, J.T. (1985). Sensation seeking, augmenting-reducing and the perceived and preferred effects of drugs. *Journal of Personality and Social Psychology, 48*, 99-106.

Kohn, P.M., Hunt, R.W. & Hoffman, F.M. (1982). Aspects of expe-rience seeking. *Canadian Journal of Behavioral Science, 14*, 13-23.

Kohn, P.M., Hunt, R.W., Davis, C.A. & Cowles, M.P. (1982). Volunteering in principle, volunteering in fact, and experience seeking. *Psychological Record, 32*, 205-213.

Kozlowski, C. (1977). Demand for stimulation and probability preferences in gambling decisions. *Polish Psychological Bulletin, 8*, 67-73.

Kramer, A.F., Sirevaag, E.J. & Braune, R. (1987). A psychophysiologi-cal assessment of operator workload during simulated flight missions. *Human Factors, 29*, 145-160.

Kretschmer, E. (1944). *Körperbau und Charakter.* Berlin: Springer.

Kruglanski, A.W. (1975). The endogenous-exogenous partition in attribution theory. *Psychological Review, 82*, 387-406.

Kruglanski, A.W., Riter, A., Amitai, A., Margolia, B.-S., Shabtai, L. & Zacks, D. (1975). Can money enhance intrinsic motivation? A test of the content-consequence hypothesis. *Journal of Personality and Social Psychology, 31*, 744-750.

Kruglanski, A.W., Riter, A., Arazi, D., Agassi, R., Montequito, J.,Peri, I. & Peretz, M. (1975). Effects of task intrinsic rewards upon extrinsic and intrinsic motivation. *Journal of Personality and Social Psychology, 31*, 699-750.

Kruse, D.L. & McBeath, A.A. (1980). Bicycle accidents and injuries: A random survey of a college population. *American Journal of Sports Medicine, 8*, 342-344.

Kuhl, J. (1978). Standard setting and risk-preference: An elaboration of the theory of achievement motivation and an empirical test. *Psychological Review, 85*, 239-248.

Kuhl, J. & Blankenship, V. (1979). The dynamic theory of achievement motivation. *Psychological Review, 86*, 141-151.

Kunkel, E. (1983). Driver improvement courses for drinking-driving reconsidered. *Accident Analysis and Prevention, 15*, 429-439.

Kunst-Wilson, W.R. & Zajonc, R.B. (1980). Affective discrimination of stimuli that cannot be recognized. *Science, 207*, 557-558.

Kuriychuk, M. (1989). *The Assessment of Psychopathy and Risk-Taking Behavior*. Unpublished Ph.D. Thesis. Queen's University Kingston, Ontario, Canada.

Lamm, R., Choueiri, E.M. & Klöckner, J.H. (1985). Accidents in the U.S. and Europe: 1970-1980. *Accident Analysis and Prevention, 17*, 429-438.

Lang, P.J. (1988). What are the Data of Emotion? In: V. Hamilton, G.H. Bower & N.H. Frijda (pp.173-193). *Cognitive Perspectives on Emotion and Motivation*. Kluwer: Dordrecht.

Langer, E.J. (1975). The illusion of control. *Journal of Personality and Social Psychology, 32*, 311-328.

Langer, E.J. (1977). The psychology of chance. *Journal of the Theory of Social Behavior, 7*, 185-207.

Langer, E.J. & Roth, J. (1975). Heads I win, Tails it's chance: The illusion of control as a function of the sequence of outcomes in a purely chance task. *Journal of Personality and Social Psychology, 32*, 951-955.

Langer, E.J. & Rodin, J. (1976). The effects of choice and enhanced personal responsibility for the aged: A field experiment in an institutional setting. *Journal of Personality and Social Psychology, 34*, 191-198.

Larrick, R.P. (1993). Motivational factors in decision theories: The role of self-protection. *Psychological Bulletin*, *113*, 440-450.

Larsen, R.J. (1987. The stability of mood variability: A spectral analytic approach to daily mood assessments. *Journal of Personality and Social Psychology*, *52*, 1195-1204.

Larsen, R.J. & Ketelaar, T. (1991). Personality and Susceptibility to Positive and Negative Emotional States. *Journal of Personality and Social Psychology, 61*, 132-140.

Larsen, R.J., Diener, E. & Emmons, R.A. (1986). Affect intensity and reactions to daily life events. *Journal of Personality and Social Psychology*, *51*, 803-814.

Larsen, R.J. & Diener E. (1987). Affect intensity as an individual difference characteristic: A review. *Journal of Research in Personality*, *21*, 1-39.

Lave, L.B. (1987). Injury as externality: An economic perspective on trauma. *Accident Analysis and Prevention*, *19*, 29-37.

Lazarus, R.S.(1991). Progress on A Cognitive-Motivational-Relational Theory of Emotion. *American Psychologist, 46*, 819-834.

Lazarus, R.S. (1984). On the Primacy of Cognition. *American Psychologist*, *39*, 124-129.

Lazarus, R.S., DeLongis, A., Folkman, S. & Gruen, R. (1985). Stress and adaptational outcomes: The problem of confounded measures. *American Psychologist*, *40*, 770-779.

Lazarus, R.S. & Folkman, S. (1984). *Stress, Appraisal, and Coping.* New York: Springer.

Lazarus, R.S. & Launier, R. (1978). Stress-related transactions between person and environment. In: L.A. Pervin & M. Lewis (Eds.). *Perspectives in Interactional Psychology*. New York: Plenum Press.

Leaman, A. & Finch, M. (1987). Impulsiveness and Venturesomness in young motorcyclists. *Personality and Individual Differences*, *8*, 945-946.

Lefcourt, H.M. (1973). The function of the illusion of control and freedom. *American Psychologist*, *28*, 417-425.

Leibnitz, G.W. (1765/1949). *New Essays Concerning Human Understanding*. (A.G. Langley, translator). La Salle, Ill.: Open Court Publishing.

Lemke, M. (1982). Correlation between EEG and driver's actions during prolonged driving under monotonous conditions. *Accident Analysis & Prevention*, *14*, 7-17.

Lesnik-Oberstein, M. & Cohen, L. (1984). Cognitive style, sensation seeking, and assortative mating. *Journal of Personality and Social Psychology*, *46*, 112-117.

Levin, B.H. & Schalmo, G.B. (1974). Self-rated liberalism is correlated with sensation seeking. *Psychological reports*, *34*, 298.
Lewicki, P. (1982). Trait relationships: The non-conscious generalization of social experiences. *Personality and Social Psychology Bulletin*, *8*, 439-445.
Lewicki, P. (1985). Nonconscious biasing effects of single instances on subsequent judgments. *Journal of Personality and Social Psychology*, *48*, 563-574.
Lewicki, P., & Hill, T. (1987). Unconscious processes as explanations of behavior in cognitive, personality, and social psychology. *Personality and Social Psychology Bulletin*, *13*, 355-362.
Lewicki, P., Hill, T., & Czyzewska, M. (1992). Nonconscious Acquisition of Information. *American Psychologist, 47*, 796-801.
Lewin, K. (1938). *The conceptual representation and the measurement of psychological forces*. Durham, N.C.: Duke University Press.
Lewin, K. (1951). In: D. Cartwright (Ed.). *Field Theory in Social Science: Selected Theoretical Papers*. New York: Harper & Row.
Libby, R. & Fishburn, P. C. (1977). Behavioral models of risk taking in business decisions: A survey and evaluation. *Journal of Accounting Research*, *1977*, 272-292.
Lichtenstein, S., Slovic, P., Fischhoff, B. & Combs, B. (1978). Judged frequency of lethal events. *Journal of Experimental Psychology: Human Learning and Memory*, *4*, 551-578.
Liebowitz, M.R. (1983) *The Chemistry of Love*. Boston: Little Brown.
Lightman, A. (1983). Weighing the Odds. *Science*, *83,4*, 21-22.
Lindeman, S.T., Van den Brink, W.P. & Hoogstraten, J. (1988). Effect of feedback on base-rate utilization. *Perceptual and Motor Skills*, *67*, 343-350.
Lindsley, D.B. (1957). Psychophysiology and Motivation. In: M.R.Jones (Ed.). *Nebraska Symposium on Motivation*, *5*. Lincoln: University of Nebraska Press.
Loehlin, J.C., Willerman, L. & Horn, J.M. (1987). Personality resemblance in adoptive families: A 10-year follow-up. *Journal of Personality and Social Psychology, 53*, 961-969.
Loehlin, J.C., Willerman, L. & Horn, J.M. (1988). Human Behavior Genetics. *Annual Review of Psychology, 39*, 101-133.
Loftus, E.F., Klinger, M.R. (1992). Is the Unconsciousness Smart or Dumb? *American Psychologist, 47,* 761-765.
Looft, W.R. (1971). Conservatives, liberals, radicals, and sensation seekers. *Perceptual and Motor Skills*, *32*, 98.

Lolas, F. Camposano, S. & Etcheberrigaray, R. (1989). Augmenting/Reducing and personality: a psychometric and evoked potential study in a Chilean sample. *Personality and Individual Differences 10*, 1173-1176.
Lopes, L.L. (1981). Decision making in the short run. *Journal of Experimental Psychology: Human Learning and Memory*, *7*, 377-385.
Lopes, L.L. (1983). Some thoughts on the psychological concept of risk. *Journal of Experimental Psychology: Human Perception and Performance*, *9*, 137-144.
Lopes, L.L. (1984). Risk and distributional inequality. *Journal of Experimental Psychology: Human Perception and Performance*, *10*, 461-485.
Lopes, L.L. (1987). Between hope and fear: the psychology of risk. *Advances in Experimental Social Psychology*, *20*, 255-295.
Lopes, L.L. & Ekberg, P-H.S. (1980). Test of an ordering hypothesis in risky decision making. *Acta Psychologica*, *45*, 161-167.
Lorenz, K. (1935). Der Kumpan in der Umwelt des Vogels. *Journal für Ornithologie*, *83*, 533-547.
Lund, A.K. & O'Neill, B. (1986). Perceived risk and driving behavior. *Accident Analysis and Prevention*, *18*, 367-370.
Lund, A.K. & Williams, A.F. (1985). A review of the literature evaluating the defensive driving course. *Accident Analysis and Prevention*, *17*, 449-460.
Lund, A.K. & Zador, P. (1984). Mandatory belt use and driver risk taking. *Risk Analysis*, *4*, 41-53.
Lykken, D.T. (1982). Fearlessness. *Psychology Today*, *16*, 20-28.
Lykken, D.T. & Tellegen, A. (1993). Is human mating adventitious or the result of lawful choice? A twin study of mate selection. *Journal of Personality and Social Psychology*, *65*, 56-68.
Lykken, D.T., Bouchard, Jr. T.J., Mc Gue, M. & Tellegen, A. (1993). Heritability of interests: A twin study. *Journal of Personality and Social Psychology*, 78, 649-661.
Lykken, D.T., Mc Gue, M. Tellegen, A. & Bouchard, Jr. T.J. (1992). Emergenesis. *American Psychologist*, *47*, 1565-1577.
MacGregor D.G. & Slovic, P. (1989). Perception of Risk in Automotive Systems. *Human Factors*, *31*, 377-389.
Machlis, G.E. & Rosa, E.A. (1990). Desired Risk: Broadening the Social Amplification of Risk Framework. *Risk Analysis*, *10*, 161-168.
Mahalel, D. & Szternfeld, Z. (1986). Safety improvements and driver perception. *Accident Analysis and Prevention*, *18*, 37-42.
Maier, S.F., Rappaport, P. & Wheatley, K.L. (1976). Conditioned inhibition and the UCS-CS interval. *Animal Learning and Behavior*, *4*, 217-220.

Malaterre, G., Ferrandez,F., Fleury, D. & Lechner, D. (1988). Decision making in emergency situations. *Ergonomics*, *31*, 643-655.
Mandler, G. (1975). *Mind and Emotion*. New York: Wiley.
Mandler, G. (1979). Thought processes, consciousness and stress. In: Hamilton, V. & Warburton, D.M. (Eds.) *Human stress and cognition* (pp.179-204). Chichester: John Wiley & Sons.
Mandler, G. & Nakamura, Y. (1987). Aspects of consciousness. *Personality and Social Psychology Bulletin*, *13*, 299-313.
Mandler, J.M. (1984). *Stories, scripts, and scenes: Aspects of schema theory*. Hillsdale, N.J.: Erlbaum.
Mann, L. & Janis, I. (1982). *Conflict theory of decision making and the expectancy - value approach*. In: N.T. Feather (Ed.). *Expectations and Actions* (pp. 341-364). Hillsdale: Lawrence Erlbaum.
March, J.G. & Shapira, Z. (1992). Variable risk preferences and the focus of attention. *Psychological Review*, *99*, 172-183.
Marshall, G.N. (1991). A multidimensional analysis of internal health locus of control beliefs: seperating the wheat from the chaff? *Journal of Personality and Social Psychology*, *61*, 483-491.
Martin, M. (1985). Individual differences in sensation seeking and atten tional ability. *Personality and Individual Differences*, *6*, 637-639.
Martin, R.A., Kuiper, N.A., Olinger, L.J. & Dobbin, J. (1987). Is stress always bad? Telic versus paratelic Dominance as a stress-moderating variable. *Journal of Personality and Social Psychology*, *53*, 970-982.
Maslow, A.H. (1954). *Motivation and Personality*. New York: Harper
Maslow, A.H. (1968). *Toward a Psychology of Being* (2nd ed.). Princeton: Van Nostrand.
Matsumoto, D. & Sanders, M. (1988). Emotional experiences during engagement in intrinsically and extrinsically motivated tasks. *Motivation and Emotion*, *12*, 353-369.
Matthews, G. (1985). The effects of extraversion and arousal on intelligence test performance. *British Journal of Psychology*, *76*, 479-493.
Matthews, G. (1986). The interactive effects of extraversion and arousal on performance: are creativity tests anomalous? *Personality and Individual Differences*, *7*, 751-761.
Matthews, G. (1987). Personality and multidimensional arousal: A study of two dimensions of extraversion. *Personality and Individual Differences*, *8*, 9-16.
Matthews, M.L. & Moran, A.R. (1986). Age differences in male drivers perception of accident risk: The role of perceived driving ability. *Accident Analysis and Prevention*, *18*, 299-313.

Matthews, G., Jones, D.M. & Chamberlain, A.G. (1989). Interactive effects of extraversion and arousal on attentional task performance: multiple resources or encoding processes. *Journal of Personality and Social Psychology*, *56*, 629-639.
Mauro, R. (1988). Opponent processes in human emotions? An experimental investigation of hedonic contrast and affective interactions. *Motivation and Emotion*, *12*, 333-351.
Mauro, R., Sato, K. & Tucker, J. (1992). The Role of Appraisal in Human Emotions: A Cross Cultural Study. *Journal of Personality and Social Psychology, 62*, 301-317.
Mayer, R.E. & Treat, J. R. (1977). Psychological, social and cognitive characteristics of high-risk drivers: a pilot study. *Accident Analysis and Prevention*, *9*, 1-8.
McClelland, D.C. (1951). *Personality*. New York: Sloane, Dryden, Holt.
McClelland, D.C. (1958). Risk taking in children with high and low need for achievement. In J.W. Atkinson (Ed.). *Motives in fantasy, action, and society* (pp. 306-321). Princeton: Van Nostrand.
McClelland, D.C. (1961). *The Achieving Society*. New York: Free Press.
McClelland, D.C. (1985). How motives, skills, and values determine what people do. *American Psychologist*, *40*, 812-825.
McDaniels, T.L. & Gregory, R.S. (1991). A framework for structuring cross-cultural research in risk and decision making. *Journal of Cross-Cultural Psychology*, *22*, 103-128.
McDougall, W. (1923). *Outline of Psychology*. New York: Scribner's.
McGrath, J.E. (1976). Stress and behavior in organizations. In: M.D. Dunnette (Ed.). *Handbook of Industrial and Organizational Psychology* (pp. 1351-1395). Chicago: Rand McNally.
McGue, M. & Bouchard, Jr. T.J. (1988). Genetic and environmental deter minants of information processing and special mental abilities: a twin analysis. In: R.J.Sternberg (Ed). *Advances in the Psychology of Human Intelligence*. Vol. *5*.
McGuire, F.L. (1976). Personality factors in highway accidents. *Human Factors*, *18*, 433-441.
McKenna, F.P. (1982). The human factor in driving accidents: an overview of approaches and problems. *Ergonomics*, *25*, 867-877.
McKenna, F.P. (1983). Accident proneness: a conceptual analysis. *Accident Analysis and Prevention*, *15*, 65-71.
McKenna, F.P. (1988). What role should the concept of risk play in theories of accident involvement? *Ergonomics*, *31*, 469-484.

McKenna, F.P. (1990). In defence of conventional safety measures: A reply to G.J.S. Wilde. *Journal of Occupational Accidents*, *11*, 171-181.
McMillan, T.M., & Rachman, S.J. (1988). Fearlessness and courage in paratroopers undergoing training. *Personality and Individual Differences*, *9*, 373-378.
McMillen, D.L., & Wells-Parker, E. (1987). The effect of alcohol consumption on risk-taking while driving. *Addictive Behaviors*, *12*, 241-247.
McMillen, D.L., Smith, S.M. & Wells-Parker, E. (1989). The effects of alcohol, expectancy, and sensation seeking on driving risk-taking. *Addictive Behaviors*, *14*, 477-483.
McNeil, B.J., Pauker, S.G., Sox, H.C. & Tversky, A. (1982). On the elicitation of preferences for alternative therapies. *New England Journal of Medicine*, *306*, 1259-1262.
Mednick, S.A. & Hutchings, B. (1978). Genetic and psycho-physiological factors in asocial behavior. In: R.D. Hare & D. Schalling, (Eds.). *Psychopathic Behavior*. New York: Wiley.
Mednick, S.A., Gabrielli, Jr. W.F. & Hutchings, B. (1987). Genetic factors in the etiology of criminal behavior. In: S.A Mednick, T.E. Moffitt & S.A. Stack (Eds.). *The causes of crime: New biological approaches* (pp.74-91). Cambridge, England: Cambridge University Press.
Melzack, R. & Casey, K.L. (1970). *The Affective Dimension of Pain*. In: M. B. Arnold (Ed.). *Feelings on Emotions*. New York: Academic Press.
Menkes, J. (1981). Risk or Angst? *Risk Analysis*, *4*, 237-240.
Mercer, G.W. (1989). Traffic accidents and convictions: group totals versus rate per kilometer driven. *Risk Analysis*, *9*, 71-77.
Merikle, P.M. (1992). Perception without Awareness: Critical Issues. *American Psychologist*, *47*, 792-795.
Metee, D.R. (1971). Rejection of unexpected success as a function of the negative consequences of accepting success. *Journal of Personality and Social Psychology*, *17*, 332-341.
Meyer, G.J. (1989). Structural convergence of mood and personality: Evidence for old and new directions. *Journal of Personality and Social Psychology*, *57*, 691-706.
Meyer, G.J. & Shack, J.R. (1989). The Structural Convergence of Mood and Personality: Evidence for Old and New "Directions". *Journal of Personality and Social Psyychology*, *57*, 691-706.
Meyer, H.H., Walker, W.B. & Litwin, G.H. (1961). Motive patterns and risk preferences associated with entrepreneurship. *Journal of Abnormal and Social Psychology*, *63*, 570-574.

Meyer, W.U., Folkes, V.S., & Weiner, B. (1976). The perceived informational value and affective consequences of choice behavior and intermediate difficulty in task selection. *Journal of Research in Personality*, *10*, 410-423.

Meyers, H.H. & Siegel, P. (1985). The motivation to breastfeed: A fit to the opponent-process theory? *Journal of Personality and Social Psychology*, *49*, 188-193.

Michaels, R.M. (1969). Tension responses of drivers generated on urban streets. *Highway Research Board Bulletin, Jan.*, 29-41.

Michon, J.A. (1989). Explanations, pitfalls and rule-based driver models. *Accident Analysis and Prevention*, *21*, 341-353.

Miller, N.E. (1948). Studies of fear as an acquired drive. I. Fear as motiva tion and fear-reduction as reinforcement in the learning of new responses. *Journal of Experimental Psychology*, *38*, 89-101.

Miller, D. P. & Swain, A. D. (1987). Human error and human reliability. In: G. Salvendy (Ed.). *Handbook of Human Factors* (pp. 219-250). New York: John Wiley & Sons.

Miller, S.B. & Ditto, B. (1989). Individual differences in heart rate and peripheral vascular responses to an extended aversive task. *Psychophysiology*, *26*, 506-513.

Miller, W.R. & Seligman, M.E.P. (1975). Depression and learned helplessness in man. *Journal of Abnormal Psychology*, *84*, 228-238.

Mineka, S. & Henderson, R.W. (1985). Controllability and predictability in acquired motivation. *Annual Review of Psychology*, *36*, 495-529.

Mischel, W. (1981). *Introduction to Personality*. New York: Holt Rinehart & Winston.

Mitchell, T.R. (1982). *Expectancy-Value Models in Organizational Psychology*. In: N.T. Feather (Ed.). *Expectations and Actions* (pp. 293-315). Hillsdale: Lawrence Erlbaum.

Mohr, D.L. & Clemmer, D.I. (1988). The "accident prone" worker: an example from heavy industry. *Accident Analysis and Prevention*, *20*, 123-127.

Mongrain, S. & Standing, L. (1989). Impairment of cognition, risk-taking and self-perception by alcohol. *Perceptual and Motor Skills*, *69*, 199-210.

Montag, I. & Comrey, A.L. (1987). Internality and externality as correlates of involvement in fatal driving accidents. *Journal of Applied Psychology*, *72*, 339-343.

Moray, N. (Ed.), (1979). *Mental Workload*. NewYork: Plenum Press.

Morrison, S.A. (1986). *Timing Skill and Risk Taking Tendencies as a Function of Exposure to Stimulating or Relaxing Music*. B.A. Thesis. Queen's University, Kingston, Ontario, Canada.
Mulder, G. (1979) Sinusarrhythmia and mental workload. In: N. Moray (Ed.). *Mental Workload* (pp.327-344). New York: Plenum Press.
Murgatroyd, S., Rushton, C. Apter, M.J. & Ray, C. (1978). The development of the Telic Dominance Scale. *Journal of Personality Assessment, 42*, 519-528.
Murgatroyd, S. (1985). The nature of telic dominance. In M.J. Apter, D. Fontana & S. Murgatroyd (Eds.). *Reversal Theory: Applications and Developments* (pp. 20-41). Cardiff, Wales: University College Press.
Muroya, H. & Ikeda, M. (1982). Mental effects of risky behavior seen by means of a portable heart rate analyzer. *Ergonomics, 25*, 476.
Murphy, D.L. & Weiss, R. (1972). Reduced monoamine oxidase activity in blood platelets from bipolar depressed patients. *American Journal of Psychiatry, 128*, 1351-1357.
Murphy, D.L., Belmaker, R. & Wyatt, R.J. (1974). Monoaminooxidase in schizophrenics and other bipolar disorders. *Journal of Psychiatric Research, 11*, 221-247.
Murphy, D.L., Belmaker, R.H., Buchsbaum, M., Martin, M.F., Ciaranello, R. & Wyatt, R.J. (1977). Biogenic amine-related enzymes and personality variations in normals. *Psychological Medicine, 7*,149-157.
Murphy, L.R., Du Bois, D. & Hurrell, J.J. (1986). Accident reduction through stress management. *Journal of Business and Psychology, 1*, 5-18.
Murphy, S.T. & Zajonc, R.B. (1993). Affect, Cognition and Awareness: Affective Priming With Optimal and Suboptimal Stimulus Exposure. *Journal of Personality and Social Psychology, 64*, 723-7739.
Murray, D.J. (1988). *A History of Western Psychology* (2nd ed.). Englewood Cliffs, N.J.: Prentice-Hall.
Murray, H.A. (1938). *Explorations in Personality*. London & New York: Oxford University Press.
Musahl, H.P. & Müller-Gethmann, H. & Thomas, C. (1991). Sind gute Wege gefährlich? Zur Gefahrenkognition bei Fahrungsunfällen im Bergbau.In: B. Zimolong & R. Trimpop (pp. 256-266). *Psychologie der Arbeitssicherheit.* Heidelberg: Asanger.
Musahl, H.P. & Alsleben, K. (1991). Zur Änderung der Gefahrenkognition: Maßnahmen und deren Evaluation. In: B. Zimolong & R. Trimpop (pp. 593-605). *Psychologie der Arbeitssicherheit.* Heidelberg: Asanger.

Musolino, R.F. & Hershenson, D.B. (1977). Avocational sensation seeking in high and low risk-taking occupations. *Journal of Vocational Behavior, 10*, 358-365.
Näätänen, R. & Summala, H. (1974). A model for the role of motivational factors in driver's decision-making. *Accident Analysis and Prevention, 6*, 243-261.
Näätänen, R. & Summala, H. (1975). A simple method for simulating danger-related aspects of behavior in hazardous activities. *Accident Analysis and Prevention, 7*, 63-70.
Näätänen, R. & Summala, H. (1976). *Road User Behavior and Traffic Accidents*. Amsterdam: North Holland.
Namdaran, F. & Elton, R.A. (1988). A study of reported injury accidents among novice motorcycle riders in a Scottish region. *Accident Analysis and Prevention, 20*, 117-121.
National Safety Council (1985). *Accident facts*. Chicago.
Neary, R.S. & Zuckerman, M. (1976). Sensation seeking, trait and state anxiety, and the electrodermal orienting response. *Psychophysiology, 13*, 205-211.
Nebylitsyn, V.D. (1972). The problem of general and partial properties of the nervous system. In: V.D. Nebylitsyn & J.A. Gray, (Eds.). *Biological Bases of Individual Behavior*. NewYork: Academic Press.
Nebylitsyn, V.D. (1972). *Fundamental Properties of the Human Nervous System*. New York: Plenum Press.
Neiss, R. (1988). Reconceptualizing arousal: Psychobiological states in motor performance. *Psychological Bulletin, 103*, 345-366.
Neiss, R. (1990). Ending arousal's reign of error: A reply to Anderson. *Psychological Bulletin, 107*, 101-105.
Neumann, P.J. & Politser, P.E. (1992). Risk and Optimality. In F.J. Yates (Ed.). *Risk Taking Behavior*. Chichester, UK: Wiley & Sons.
Newcomb, M.D. & McGee, L. (1991). Influence of Sensation Seeking on General Deviance and Specific Problem Behaviors From Adolescence to Young Adulthood. *Journal of Personality and Social Psychologie, 61*, 614-628.
Newman, S.E. & Benassi, V.A. (1989). Putting judgements of control into context: Contrast effects. *Journal of Personality and Social Psychology, 56*, 876-889.
Nicholls, J.G. (1984). Achievement motivation: Conceptions of ability, subjective experience, task choice, and performance. *Psychological Review, 91*, 328-346.

Nichols, R.C. (1978). Twin studies of ability, personality, and interests. *Homo, 29*, 158-173.
Nicholson, A.N., Hill, L.E., Borland, R.G. & Ferres, H.M. (1970). Activity of the nervous system during the let-down, approach and landing: A study of short duration high workload. *Aerospace Medicine, 41*, 436-446.
Nicholson, A.N., Hill, L.E., Borland, R.G. & Krzanowski, W.J. (1973). Influence of workload on the neurological state of a pilot during approach and landing. *Aerospace Medicine, 44*, 146-152.
Noordzij, P.C., Meester, A.C. & Verschuur, W.L.G. (1988). Night time driving: the use of seat-belts and alcohol. *Ergonomics, 31*, 663-668.
O'Connor, K., Hallam, R. & Rachman, S. (1985). Fearlessness and courage. *British Journal of Psychology, 76*, 187-197.
OECD (1990). *Behavioural adaptations to changes in the road transport system*. OECD: Paris, 1990.
Ogden, G. D., Levine, J. M. & Eisner, E. (1979). Measurement of workload by secondary task. *Human Factors,21*,529-548.
Olds, J. & Olds, M. (1965). Drives, rewards, and the brain. In: *New Directions in Psychology, II*, New York: Holt, Rhinehart & Winston.
Olds, M.E. & Fobes, J.L. (1981). The central basis of motivation: Intracranial self-stimulation studies. *Annual Review of Psychology, 32*, 523-574.
Oliver, J.E. (1985). Successive generations of child maltreatment: Social and medical disorders in the parents. *British Journal of Psychiatry, 147*, 484-490.
Oliver, J.E. (1988). Successive generations of child maltreatment: The children. *British Journal of Psychiatry, 153*, 543-553.
Ontario Road Safety *Annual Report 1987* (1988).
Oppe, S. (1988). The concept of risk: a decision theoretic approach. *Ergonomics, 31*, 435-440.
Oppenheim, L. (1977). *Ancient Mesopotamia*. University of Chicago Press.
Ore, O. (1960). Pascal and the invention of probability theory. *American Mathematical Monthly, 67*, 409-419.
Orr, L. (1982). Goals, Risks and Choices. *Risk Analysis, 2*, 239-241.
Ortega, D.F. & Pipal, J.E. (1984). Challenge seeking and the Type A coronary-prone behavior pattern. *Journal of Personality and Social Psychology, 46*, 1328-1334.
Otway, H.J. & Cohen, J.J. (1975). Revealed Preferences: comments on the Starr benefit-risk relationship. *RM-75-5*, International Institute for Applied Systems Analysis, Laxenburg, Austria.

Otway, H.J., & Fishbein, M. (1977). Public attitudes and decision making. *RM-77-54*, International Institute for Applied Systems Analysis, Laxenburg, Austria.

Otway, H.J. & von Winterfeldt, D. (1982). Beyond acceptable risk: On the social acceptability of technologies. *Policy Sciences, 14*, 247-256.

Oxendine, J.B. (1970). Emotional arousal and motor performance. *Quest, 13*, 23-32.

Paisey, T.J.H. & Mangan, G.L. (1980). The relationship of extraversion, neuroticism, and sensation-seeking to questionnaire-derived measures of nervous system properties. *Pavlovian Journal of Biological Sciences, 15*, 123-130.

Panksepp, J. (1986). The neurochemistry of behavior. *Annual Review of Psychology, 37*, 77-107.

Parker, L.E. & Lepper, M.R. (1992). Effects on Fantasy Contexts on Children's Learning and Motivatiion: Making Learning More Fun. *Jornal of Personality and Social Psychology, 62*, 625-633.ß

Parkes, L. (1987). *A Cross-Situational Investigation of Risk-Taking Disposition*. B.A. Thesis. Queen's University, Kingston, Ontario, Canada.

Partyka, S.C. (1984). Simple models of fatality trends using employment and population data. *Accident Analysis and Prevention, 16*, 211-222.

Paterson, R.J. & Neufeld, R.W.J. (1987). Clear danger: Situational deter minants of the appraisal of threat. *Psychological Bulletin,101*,404-416

Pavlov, I.P. (1927) *Conditioned Reflexes*. (G.V. Anrep, translator, 1960). New York: Dover.

Pavlov, I.P. (1955) *Selected Works*. (S. Belsky, translator) Moscow: Foreign Langauge Publishing House.

Payne, J. (1976). Task Complexity and Contingent Information Processing in Decision Making: An Information Search and Protocol Analysis, *Organizational Behavior and Human performance, 16*, 366-387.

Peltzman, S. (1975). The effects of automobile safety regulation. *Journal of Political Economy, 83*, 677-725.

Pelz, D.C. & Krupat, E. (1974). Caution profile and driving record of undergraduate males. *Accident Analysis and Prevention, 6*, 45-58.

Pelz, D.C. & Schumann, S.H. (1971). Are young drivers really more dangerous after controlling for exposure and experience? *Journal of Safety Research, 3*, 68-79.

Perez, J. & Torrubia, R. (1985). Sensation seeking and antisocial behavior in a student sample. *Personality and Individual Differences, 6*, 401-403.

Perrow, C. (1984) *Normal Accidents: Living with High-Risk Technologies*. New York: Basic Books.

Petras, A.F. & Hoffman, E.P. (1983). Roentgenographic skeletal injury patterns in parachute jumping. *American Journal of Sports Medicine, 11*, 325-328.
Pfaff, M., Kistler, E., Schulze, H. & Theis, A. (1982). *Methoden und Kriterien zur Überprüfung des Erfolgs von Aufklärungskampagnen*. Forschungsbericht. Köln: Bundesanstalt für Straßenwesen, Bereich Unfallforschung (Hrsg.).
Pfafferott, I. & Huguenin, R.D. (1993). Adaptation nach Einführung von Sicherheitsmaßnahmen. *Zeitschrift für Verkehrssicherheit, 37*, 71-83.
Pidgeon, N.F. (1991). Safety Culture and Risk Management in Organizations. *Journal of Cross-Cultural Psychology, 22*, 129-140.
Piët, S. (1987). What motivates stunt men? *Motivation and Emotion, 11*, 195-213.
Piliavin, J., Callero, P. & Evans, D. (1982). Addiction to altruism? Opponent-process theory and habitual blood donation. *Journal of Personality and Social Psychology*, *43*, 1200-1213.
Platenius, P.H. & Wilde, G.J.S. (1989). Personal characteristics related to accident histories of Canadian pilots. *Aviation, Space, and Environmental Medicine*, *60*, 42-45.
Plomin, R. & Rende, R. (1991). Human behavior genetics. *Annual Review of Psychology*, *42*, 161-190.
Pomerleau, O. (1979). Why people smoke: Current psychobiological models. In: P. Davidson (Ed.). *Behavioral medicine: Changing health lifestyles*. New York: Brunner / Mazel.
Popper, K. (1974). Replies to my critics. In: P.A. Schillp (Ed.). *The philosophy of Karl Popper*, La Salle: Open Court.
Poulton, E.C. (1970). *Environment and Human Efficiency*. Springfield: Thomas
Povinelli, D.J. (1993). Reconstructing the evolution of mind. *American Psychologist*, *48*, 493-509.
Premack, D. (1976). *Intelligence in ape and man*. Hillsdale:Erlbaum.
Pribram, K.H. & McGuiness, D. (1975). Arousal, activation, and effort in the control of attention. *Psychological Review*, *2*, 116-149.
Pruitt, D.G. & Kimmel, M.J. (1977). Twenty years of experimental gaming: critique, synthesis, and suggestions for the future. *Annual Review of Psychology*, *28*, 363-392.
Quadrel, M.J., Fischhoff, B. & Davis, W. (1993). Adolescent (In)vulnerability. *American Psychologist*, *48*, 102-116.

Rabbitt, P. (1979). Current paradigms and models in human information processing. In: V. Hamilton, & D.M. Warburton (Eds.). *Human stress and cognition*. Chichester: John Wiley & Sons.

Ranieri, D. & Zeiss, A. (1984). Induction of depressed mood: A test of opponent-process theory. *Journal of Personality and Social Psychology*, *47*, 1413-1422.

Rasmussen, J. (1982), Human errors. A taxonomy for describing Human Malfunction in industrial Installations. *Journal of Occupational Accidents*, *4*, 311-333.

Rasmussen, J. (1985). Trends in human reliability analysis. *Ergonomics*, *28*, 1185-1195.

Rawlings, D. (1984). The correlation of EPQ Psychoticism with two beha vioral measures of impulisivity. *Personality and Individual Differences*, *5*, 591-594.

Rawson, A.J. (1944). Accident proneness. *Psychosomatic Medicine*, *6*, 88-94.

Redlich, F. (1957). Towards a better theory of risk. *Explorations in Entre preneurial History*, *10*, 33-39. Cited in: Schäfer, R.E. (1978). *What are we talking about when we talk about "Risk"? A critical survey of risk and risk preference theories*. RM-78-69, International Institute for Applied Systems Analysis, Laxenburg, Austria.

Reisenzein, R. & Schönpflug, W. (1992). Stumpf's Cognitive-Evaluative Theory of Emotion. *American Psychologist, 47*, 34-45.

Renn, O. (1981). *Man, Technology and Risk: A study on intuitive risk assessment and attitudes towards nuclear energy*. Spezielle Berichte der Kernforschungsanlage Jülich, *115*.

Rheinberg, F. (1991). Flow experience when motorcycling: a study of a special human condition. In R. Brendicke (Ed.), *Safety Environment Future*. Proceedings of the 1991 International motorcycle conference. Forschungsheft 7 (350-362). Bochum:Institut für Zweiradsicherheit.

Rheinberg, F., Dirksen, U. & Nagels, E. (1986). Motivationsanalysen zu verschieden riskantem Motorradfahren. *Zeitschrift für Verkehrssicher-heit*, *32*, 75-80.

Rissler, A. (1977). Stress reactions at work and after work during a period of quantitative overload. *Ergonomics*, *20*, 13-16.

Roberts, J.M., Thompson, W.E. & Sutton-Smith, B. (1966). Expressive self-testing in driving. *Human Organization*, *25*, 54-63.

Robinson, D.W. (1985). Stress seeking: Selected behavioral characteristics of elite rock climbers. *Journal of Sport Psychology*, *7*, 400-404.

Robinson, P. (1961). *The Measurement of Achievement Motivation.* Unpublished doctoral dissertation, University of Oxford. Cited in:Murgatroyd, S., Rushton, C. Apter, M.J. & Ray, C. (1978). The development of the Telic Dominance Scale. *Journal of Personality Assessment,42*,519-528.
Rofè, Y. (1984). Stress and affiliation: a utility theory. *Psychological Review, 91*, 235-250.
Rogers, T.B. (1987). Evidence for sensation seeking behavior during assessment of the trait: a note on the construct validity of the measurement operation. *Personality and Individual Differences, 8*, 957-959.
Rohrmann, B. (1992). Risiko Kommunikation, Aufgaben-Konzepte-Evaluation. In: B. Zimolong & R. Trimpop (pp.577-592). *Psychologie der Arbeitssicherheit.* Heidelberg: Asanger Verlag.
Rohrmann, B. (in press). Perception and evaluation of risk: a cross-cultural comparison.
Rossiter, C.M.E. (1987). *A Behavioral Measure of Risk Taking and its Relationship to Questionnaire Assessment of the Risk-Taking Personality, Self-Control and Personal Accident History.* B.A. Thesis. Queen's University, Kingston, Ontario, Canada.
Rothbart, M.K. (1973). Laughter in young children. *Psychological Bulletin, 80*, 247-256.
Rothbaum, F. Weisz, J.R. & Snyder, S.S. (1982). Changing the world and changing the self: A two-process model of perceived control. *Journal of Personality and Social Psychology, 42*, 5-37.
Rothengatter, T. (1988). Risk and the absence of pleasure: a motivational approach to modelling road user behavior. *Ergonomics, 31*, 599-607.
Rotter, J.B. (1966). Generalized expectancies for internal versus external control of reinforcement. *Psychological Monographs, 80*, (No.609).
Rotter, J.B. (1990). Internal versus external control of reinforcement: A case story of a variable. *American Psychologist, 45*, 489-493.
Roviaro, S.E. & Holmes, D.S. (1980). Arousal transfer: the influence of fear arousal on subsequent sexual arousal for subjects with high and low sex guilt. *Journal of Research in Personality, 14*, 307-320.
Rowland, G.L. & Franken, R.E. (1986). The four dimensions of sensation seeking: a confirmatory factor analysis. *Personality and Individual Differences, 2*, 237-240.
Rowland, G.L., Franken, R.E. & Harrison, K. (1986). Sensation seeking and participation in sporting activities. *Journal of Sport Psychology, 8*, 212-220.

Rowland, G.L., Franken, R.E., Williams, S.E. & Heatherton, T. (1988). The perception of sensation seeking in familiar and unfamiliar others. *Personality and Individual Differences*, *9*, 237-241.

Rumar, K., Bergrund, U., Jernberg, P., & Ytterbom, U. (1976). Driver reaction to a technical safety measure-studded tires. *Human Factors*, *18*, 443-454.

Russell, J.A., Weiss, A. & Mendelsohn, G.A. (1989). Affect Grid: A single- item scale of pleasure and arousal. *Journal of Personality and Social Psychology*, *57*, 493-502.

Russell, J.A. (1991). Culture and the Categorization of Emotions. *Psychological Bulletin, 110*, 426-450.

Ryan, R.M. & Grolnick, W.S. (1986). Origins and pawns in the class room: Self-report and projective assessment of individual differences in children' perception. *Journal of Personality and Social Psychology*, *50*, 736-750.

Sansone, C., Weir, C. Harpster, L. & Morgan, C. (1992). Once a boring task always a boring task? interest as a self-regulatory mechanism. *Journal of Personality and Social Psychology*, *63*, 379-390.

Sarason, I.G. (1978). The test anxiety scale: Concept and research. In: C.D. Spielberger & I.G. Sarason (Eds.). *Stress and Anxiety*, *Vol.5* (pp. 193-216). Washington, DC: Hemisphere.

Savage, L. (1972). *The Foundations of Statistics* (2nd ed.). New York: Dover.

Scarr, S. & McCartney , K. (1983). How people make their own environments: A theory of Genotype-Environment effects. *Child Development*, *54*, 424-435.

Scarr, S., Webber, P.L., Weinberg, R.A. & Wittig, M.A. (1981). Personality resemblance among adolescents and their parents in biologically related and adoptive families. *Journal of Personality and Social Psychology*, *40*, 885-898.

Schachter, S. (1970). The assumption of identity and peripheralist-centralist controversies in motivation and emotion. In: M.B. Arnold (Ed.) *Feelings and Emotion: The Loyola Symposium*.New York:Academic Press.

Schachter, S. (1971). *Emotion, Obesity and Crime*. New York: Academic Press.

Schachter, S. & Singer, J. (1962). Cognitive, social and physiological determinants of emotional state. *Psychological Review*, *69*, 378-399.

Schäfer, R.E. (1978). *What are we talking about when we talk about Risk? A critical survey of risk and risk preference theories*. RM-78-69, International Institute for Applied Systems Analysis, Laxenburg, Austria.

Schalling, D. (1978). Psychopathy-related personality variables and the psychophysiology of socialization. In: R.D. Hare & D. Schalling, (Eds.) *Psychopathic Behavior*. New York: Wiley.
Schalling, D., Edman, G. & Asberg, M. (1983). Impulsive cognitive style and inability to tolerate boredom: Psychobiological studies of tempera mental vulnerability. In: M. Zuckerman (Ed.). *Biological Bases of Sensation Seeking, Impulsivity, and Anxiety* (pp. 123-146). Hillsdale: Lawrence Erlbaum.
Schmitz, B. & Skinner, E. (1993). Perceived Control, Effort, and Academic Performance: Interindividual, Intraindividual, and Multivariate Time-Series Analyses. *Journal of Personality and Social Psychology, 64*, 1010-1028.
Schneider, K. & Posse, N. (1982). Risk taking in achievement-oriented situations: Do people really maximize affect or competence information? *Motivation and Emotion*, *6*, 259-271.
Schoemaker, P.J.H. (1982). The Expected Utility Model: It's Variance Purposes, Evidence and Limitations. *Journal of Economic Literature, 20*, 529-563.
Scholz, R.W., Seydel, U., Rechbauer, W. & Wentz, S. (1983). Zum Zusammenhang zwischen Verkehrsauffälligkeit, Persönlichkeitsfaktoren und Verhalten in Konfliktspielen. *Zeitschrift für experimentelle und angewandte Psychologie, 30,* 273-298.
Schurr, P.H. (1987). Effects of Gain and Loss Decision Frame on Risky Purchase Negotiations. *Journal of Applied Psychology*, *72*, 351-358.
Schulz, U. (1993). Freizeitmotivationen von Motorradfahrern und ihre Aus wirkungen auif die Verkehrssicherheit. *Zeitschrift für Verkehrssicherheit*, *39*, 68-81.
Schulz, U. & Kerwien, H. (1991). Zur Risikobereitschaft junger Motorradfahrer *Zeitschrift für Verkehrssicherheit*, *37*, 168-171.
Schwarz, R.M., Burkhart, B.R. & Green, S.B. (1978). Turning on or turning off: sensation seeking or tension reduction as motivational determinants of alcohol use. *Journal of Consulting and Clinical Psychology*, *46*, 1144-1145.
Schwing, R.C. & Albers Jr., W.A. (Eds.) (1980). *Societal Risk Assessment: How safe is safe enough?* New York: Plenum Press.
Segal, N.L. (1993). Twin, sibling, and adoption methods: test of a evolutionary hypotheses. *American psychologist*, *48*, 943-956.
Seligman, M.E.P. (1975). *Helplessness*. San Francisco: Freeman.

Seligman, M.E.P. & Weiss, J.M. (1980). Coping behavior: Learned helplessness, physiological change and learned inactivity. *Behavior, Research and Therapy, 18*, 459-512.

Shannon, H.S. (1986). Road accident data: Interpreting the British experience with particular reference to the risk homeostasis theory. *Ergonomics, 29*, 1005-1015.

Sharit, J. & Salvendy, G. (1982). Occupational stress: Review and Reappraisal. *Human Factors, 24*, 129-162.

Shaver, P., Schwartz, J., Kirson, D. & O'Connor, C. (1987). Emotion knowledge: Further exploration of a prototype approach. *Journal of Personality and Social Psychology, 52*, 1061-1086.

Shipley, P. (1987). The management of psychosocial risk factors in the workplace environment: UK law compared. *Work and Stress, 1*,43-48.

Short, Jr., J.F. (1984). The social fabric at risk: toward the social transformation of risk analysis. *American Sociological Review, 49*, 711-725.

Siegel, A. I., Bartter, W. D., Wolf, J. J., Knee, H. E. & Haas, P. M. (1984). *Maintenance personnel performance simulation (MAPPS) model: Summary description.* (Applied Psychological Services and Oak Ridge National Laboratory, NUREG/CR-3626). Washington DC: US Nuclear Regulatory Commission.

Siero, S. Boon, M., Kok, G. & Siero, F. (1989). Modification of driving behavior in a large transport organization: A field experiment. *Journal of Applied Psychology, 74*, 417-423.

Singleton, W.T., Fox, J. G. & Whitfield, D. (1971). *Measurement of Man at Work: An appraisal of physiological and psychological criteria in man-machine systems.* London: Taylor and Francis LTD.

Sivak, M., Soler, J. & Tränkle, U. (1989) Cross-cultural differences in driver risk-taking. *Accident Analysis and Prevention, 21*, 363-369.

Sivak, M., Soler, J. & Tränkle, U. (1989b) Cross-cultural differences in driver self-assessment. *Accident Analysis and Prevention,21*,371-375

Sivak, M., Soler, J., Tränkle, U. & Spagnhol, J.M. (1989) Cross-cultural differences in driver risk-perception. *Accident Analysis and Prevention, 21*, 355-362.

Sjöberg, L. (1980). The risks of risk analysis. *Acta Psychologica, 45*, 301-321.

Skinner, B.F. (1966). *The Behavior of Organisms.* Englewood Cliffs, N.J.: Prentice Hall.

Sleet, D.A. (1984). *Health and the psychology of restraint system use.* Paper presented at the 92nd Annual Convention of the American Psychological Association, Toronto.

Slovic, P. (1967). The relative influence of probabilities and payoffs upon perceived risk of a gamble. *Psychonomic Science, 9*, 223-224.
Slovic, P. (1978). The psychology of protective behavior. *Journal of Safety Research, 10*, 58-68.
Slovic, P. & Fischhoff, B. (1982). Targeting risk: comment on "Wilde's Theory of Risk Homeostasis". *Risk Analysis, 2*, 227-234.
Slovic, P., Fischhoff, B. & Lichtenstein, S. (1977). Behavioral decision theory. *Annual Review of Psychology, 28*, 228-267.
Slovic, P., Fischhoff, B. & Lichtenstein, S. (1980). Perceived Risk. In: R.C. Schwing & W.A. Albers Jr. (Eds.). *Societal Risk Assessment: How Safe Is Safe Enough?* New York: Plenum Press.
Slovic, P., Fischhoff, B. & Lichtenstein, S. (1982). Why study risk perception? *Risk Analysis, 2*, 83-93.
Slovic, P. & Lichtenstein, S. (1968). Relative importance of probability and payoffs in risk taking. *Journal of Experimental Psychology Monograph, 78*, no.3, part 2.
Slovic, P., MacGregor, D. & Kraus, N.N. (1987). Perception of risk from automobile safety defects. *Accident Analysis and Prevention, 19*, 359-373.
Smith, C.A. (1989). Dimensions of appraisal and physiological response in emotion. *Journal of Personality and Social Psychology,56*,339-353.
Smith, C.A. & Ellsworth, P.C. (1985). Patterns of cognitive appraisal in emotion. *Journal of Personality and Social Psychology, 48*, 813-838.
Smith, K.C. & Apter, M.J. (1975). *A Theory of Psychological Reversals.* Chippenham, U.K.: Picton Publishing.
Smith, B.D., Wilson, R.J. & Jones, B.E. (1983). Extraversion and multiple levels of caffeine-induced arousal: Effects on overhabituation and dishabituation. *Psychophysiology, 20*, 29-34.
Smith, B.D., Davidson, R.A., Smith, D.L., Goldstein, H. & Perlstein,W. (1989). Sensation seeking and arousal: effects of strong stimulation on electrodermal activation and memory task performance. *Personality and Individual Differences, 10*, 671-679.
Smith, R.E., Ptacek, J. T. & Smoll, F.L. (1992). Sensation Seeking, Stress and Adolescent Injuries: A Test of Stress-Buffering, Risk Taking, and Coping Skills Hypothesis. *Journal of Personality and Social Psychology, 62*, 1016-1024.
Solomon, R.L. & Corbit, R. (1974). An opponent-process theory of motivation: Temporal dynamics of affect. *Psychological Review, 81*, 119-145.

Solomon, R.L. (1980). The opponent-process theory of acquired motivation: The costs of pleasure and the benefits of pain. *American Psychologist*, *35*, 691-712.
Sommers, S. & Scioli, A. (1986). Emotional range and value orientation: Toward a cognitive view of emotionality. *Journal of Personality and Social Psychology*, *51*, 417-422.
Sorrentino, R.M., Hewitt, E.C. & Raso-Knott, P.A. (1992). Risk-Taking in games of chance and skill: informational and affective influences on choice behavior. *Journal of Personality and Social psychology, 62*, 522-533.
Spence, K.W. (1956). *Behavior theory and conditioning*. New Haven: Yale University Press.
Sperandio, J. C. (1978). The regulation of working methods as a function of workload among air traffic controllers. *Ergonomics*, *21*, 195-202.
Spielberger, C.D., Gorsuch, R.L. & Lushene, R.E. (1970). *Manual for the trait-state anxiety inventory*. Palo Alto: Consulting Psychologists Press.
Spielberger, C.D, & Frank, R.G. (1992). Injury Control. *American Psychologist*, *47*, 1029-1030.
Stacy, A.W., Newcomb, M.D. & Bentler, P.M. (1991). Social psychological influences on sensation seeking from adolescence to adulthood. *Personality and Social Psychology Bulletin*, *17*, 701-708.
Stacy, A.W., Newcomb, M.D. & Bentler, P.M. (1991b). Personality, problem drinking, and drunk driving: mediating, moderating and direct-effect models. *Journal of Personality and Social Psychology*, *60*, 795-811.
Starr, C. (1969). Social benefit vs. technological risk. *Science*, *165*, 1232 - 1238.
Statistics Canada 1981, (1983). *Census of Canada: Population by Age, Sex and Marital Status*. Catalogue, 92-901.
Staw, B.M. & Ross, J. (1985). Stability in the midst of change: A dispositional approach to job attitudes. *Journal of Applied Psychology*, *70*, 469-480.
Staw, B.M., Sandelands, L.E. & Dutton, J.E. (1981). Threat-rigidity effects in organizational behavior: A multi-level analysis. *Administrative Science Quarterly*, *26*, 501-524.
Stein, L. (1983). The chemistry of positive reinforcement. In: M. Zuckerman (Ed.). (1983). *Biological Bases of Sensation Seeking, Impulsivity, and Anxiety* (pp. 151-176). Hillsdale: Lawrence Erlbaum.

Stelmack, R.M., Houlihan, M. & McGarry-Roberts, P.A. (1993). personality, reaction time, and event-related potential. *Journal of Personality and Social Psychology*, *65*, 399- 409.

Stemmler, G. & Meinhardt, E. (1990). Personality, situation and physiological arousability. *Personality and Individual Differences. 11*, 293-308.

Stepper, S. & Strack, F. (1993). Proprioceptive Determinants of Emotional and Nonemotional Feelings. *Journal of Personality and Social Psychology, 64*, 211-220.

Stirling, J. (1977). *Strength of the nervous system, extraversion-introversion, and kinesthetic and cortical augmenting and reducing.* Unpublished Doctoral Dissertation, University of York, England.

Strasser, H. (1979). Measurement of mental workload. In: N. Moray (Ed.). *Mental Workload.* New York: Plenum Press.

Straub, W.F. (1982). Sensation seeking among high and low-risk male athletes. *Journal of Sport Psychology*, *4*, 246-253.

Streff, F.M. & Geller, E.S. (1988). An experimental test of risk compensation: between-subject versus within-subject analyses. *Accident Analysis and Prevention*, *20*, 277-287.

Strelau, J. (1972). The general and partial nervous system types-data and theory. In *Biological Bases of Individual Behavior.* V.D. Nebylitsyn & J.A. Gray, (Eds.) New York:Academic Press.

Strelau, J. (1974). Temperament as an expression of energy level and temporal features of behavior. *Polish Psychological Bulletin*, *5*, 119-127.

Strelau, J. (1982). Biologically Determined Dimensions of Personality or Temperament? *Personality and Individual Differences. 3*, 355-360.

Strelau, J. (1983). *Temperament, Personality, Activity.* London: Academic Press.

Strelau, J. (1984). Zuckerman's sensation seeking theory: A view from Eastern Europe. *Behavior and Brain Science*, *7*, 451-452.

Streufert, S., Pogash, R.M., Roache, J., Gingrich, D., Landis, R. Severs, W.,Lonardi, L. & Kanter, A. (1992). Effects of alcohol intoxication on risk taking strategy, and error rate in visuomotor performance. *Journal of Applied Psychology*, *77,* 515-524.

Strickland, B.R. (1990). Internal-external control expectancies: From contingency to creativity. *American Psychologist*, *44*,1-12.

Strickland, L.H., Lewicki, R.J. & Katz, A.M. (1966) Temporal orientation and perceived control as determinants of risk-taking. *Journal of Experimental Social Psychology*, *2*, 143-151.

Sulzer-Azaroff, B. (1987). The modification of occupational safety behavior. *Journal of Occupational Accidents*, *9*, 177-197.

Summala, H. (1988). Risk control is not risk adjustment: the zero-risk theory of driver behavior and its implications. *Ergonomics*, *31*, 491-506.
Svebak, S. & Stoyva, J. (1980). High arousal can be pleasant and exciting: The theory of psychological reversals. *Biofeedback and SelfRegulation*, *5*, 439-444.
Svebak, S. Storfjell, O. & Dalen, K. (1982). The effects of a threatening context upon motivation and task-induced physiological changes. *British Journal of Psychology*, *73*, 505-512.
Svebak, S. & Kerr, J.H. (1989). The role of impulsivity in preference for sports. *Personality and Individual Differences*, *10*, 51-58.
Svenson, O. (1978). Risks of road transportation in a psychological perspective. *Accident Analysis and Prevention*, *10*, 267-280.
Svenson, O. (1984). Cognitive processes in judging cumulative risks over different periods of time. *Organizational Behavior and Human Performance*, *33*, 22-41.
Svenson, O. (1989). On Expert Judgment in Safety Analyses in the Process Industries. Reliability Engineering and System Safety, 25, 219-256.
Svenson, O., Fischhoff, B. & MacGregor, D. (1985). Perceived driving safety and seatbelt usage. *Accident Analysis and Prevention*, *17*, 119-133.
Swain, A. D. & Guttmann, H. E. (1983). *Handbook of human reliability analysis with emphasis on nuclear power plant applications*. In Sandia National Laboratories, NUREG/CR-1278 (Ed.), Washington DC: US Nuclear Regulatory Commission.
Synodinos, N.E. & Papacostas, C.S. (1985). Driving habits and behavior patterns of university students. *International Review of Applied Psychology*, *34*, 241-257.
Takata, T. K. (1982) The influence of environmental noises on the performance of a repetitive task. *Ergonomics*, *25*, 470.
Taylor, D.H. (1964). Drivers' galvanic skin response and the risk of accident. *Ergonomics*, *7*, 253-262.
Taylor, D.H. (1976). Accidents, risks, and the models of explanation. *Human Factors*, *18*, 371-380.
Taylor, S.E. & Brown, J.D. (1988). Illusion and well-being: A social psychological perspective on mental health. *Psychological Bulletin*, *103*, 193-210.

Taylor, S.E., Kemeny, M.E. Aspinwall, L.G., Schneider, S.G., Rodriguez, R. & Herbert M. (1992). Optimism, coping, psychological distress and high-risk sexual behavior among men at risk for acquired immunodeficiency syndrome (AIDS). *Journal of Personality and Social Psychology, 63,* 460-473.

Tellegen, A. Lykken, D.T., Bouchard, Jr, T.J., Wilcox, K.J., Segal, N.L. & Rich, S. (1989). Personality similarity in twins reared apart and together. *Journal of Personality and Social Psychology*, (in press).

Tesser, A. (1993). The importance of heritability in psychological research. *Psychological Review*, *100*, 129-142.

Tetens, J. N. (1786). *Einleitung zur Berechnung der Leibrenten und Anwartschaften*. Leipzig.

Thackray, R.I. & Pearson, D.W. (1968). Effects of cognitive appraisal of stress on heart rate and task performance. *Perceptual and Motor Skills*, *27*, 651-658.

Thayer, R.E. (1967). Measurement of activation through self-report. *Psychological Reports*, *20*, 663-678.

Thayer, R.E. (1972). Reported activation and verbal learning as a function of group size (social facilitation) and anxiety-inducing instructions. *The Journal of Social Psychology*, *88*, 277-287.

Thayer, R.E. (1978). Factor Analytic and Reliability Studies on the Activation-Deactivation Adjective Checklist. *Psychological Reports*, *42*, 747-756.

Thayer, R.E. (1987). Problem perception, optimism, and related states as a function of time of day (diurnal rhythm) and moderate exercise: two arousal systems in interaction. *Motivation and Emotion*, *11*, 19-36.

Thayer, R.E., Takahashi, P.J. & Pauli, J.A. (1988). Multidimensional arousal states, diurnal rhythms, cognitive and social processes, and Extraversion. *Personality and Individual Differences*, *9*, 15-24.

Thieme, R.E. & Feij, J.A. (1986). Tyramine, a new clue to disinhibition and sensation seeking?. *Personality and Individual Differences*, *7*, 349-354.

Thompson, S. C. (1981). Will it hurt less if I can control it? A complex answer to a simple question. *Psychological Bulletin*, *90*, 89-101.

Thompson, S.C., Sobolew-Shubin, A., Galbraith, M.E., Schwankowsky, L. & Cruzen, D. (1993). Maintaining perceptions of control: finding perceived control in low-control circumstances. *Journal of Personality and Social Psychology*, *64*, 293-304.

Thornton, D. (1985). Rate of offending, risk-evaluation and risk-preference. *Personality and Individual Differences*, *6*, 127-128.

Tillman, W.A. & Hobbs, G.E. (1949). The accident prone automobile driver. *American Journal of Psychiatry, 106*, 321-331.
Toda, M. (1980). Emotions and decision making. *Acta Psychologica, 45*, 133-155.
Tolman, E.C. (1926). The nature of the fundamental drives. *Journal of Abnormal Psychology, 20*, 349-358.
Tolman, E.C. (1955). Principles of performance. *Psychological Review, 62*, 315-326.
Tomaka, J., Blascovitch, J., Kelsey, R.M., & Leitten, C.L. (1993). Subjective, physiological, and behavioral effects of threat and challenge appraisal. *Journal of Personality and Social Psychology, 65*, 248-260.
Tomkins, S.S. (1962). *Affect, Imagery, Consciousness: Vol.1, The Positive Affects*. New York: Springer.
Torg, J.S., Quedenfeld, T.C., Burstein, A., Spealman, A. & Nichols III, C. (1979). National football head and neck injury registry: a report on cervical quadriplegia, 1971-1975. *American Journal of Sports Medicine, 7*, 127-132.
Tränkle, U., Gelau, C. & Metker, T. (1989). Einflüsse von Alter und Geschlecht auf die Wahrnehmung situationsspezifischer Risiken im Straßenverkehr. *Zeitschrift für experimentelle und angewandte Psychologie, 36*, 311 - 327.
Tränkle, U. & Gelau, C. Maximization of subjective expected utility or risk control ? Experimental tests of risk homeostasis theory. *Ergonomics, 35*, 7-23.
Transport Canada 1983 (1984). Canadian Motor Vehicle Traffic Accidents.TP3322, Road Safety Directorate, Transport Canada, Ottawa.
Trimpop, R.M. (1988). *Psychological factors involved in accident causation and accident prevention*. Report presented to the Industrial Accident Prevention Association. Kingston, Ontario,Canada.
Trimpop, R.M. (1989). *Risk Compensation, and the Interaction of Personality and Situational Variables*. Paper presented at the 1st European Congress of Psychology, Amsterdam, The Netherlands.
Trimpop, R.M. (1989a). *Employee Perceptions about Health and Safety*. Report for Davidson Instrument Panel Safety Personnel for the Industrial Accident Prevention Association. Kingston, Ontario, Canada.
Trimpop, R.M. (1993 in press) Risk Compensation and the interaction of personality and situational variables. In: R.M. Trimpop & G.J.S. Wilde (Eds.). *Challenges to Accident Prevention: The issue of risk compensation behaviour*(pp.135-146). Groningen, The Netherlands: STYX

Trimpop, R.M. (1994 in press) Motivation zur Arbeitssicherheit, *Der Sicherheitsingenieur.*

Trivers, R. (1972). Parental Investment and Sexual selection. In: B.Campbell (Ed.) *Sexual Selection and the Descent of Man*(pp. 136-179). Chicago: Aldine-Atherton.

Trivers, R. (1985). *Social Evolution*. Menlo Park: Benjamin / Cummings.

Turkheimer, E. (1991). Individual and Group Differences in Adoption Studies of IQ. *Psychological Bulletin, 110*, 392-406.

Tversky, A. (1975). *On the Elicitation of Preferences: Descriptive and Prescriptive Considerations*. Paper presented at the workshop on decision making with multiple conflicting objectives. International Institute for Applied Systems Analysis. Laxenburg: Austria.

Tversky, A. & Kahneman, D. (1981). The framing of decisions and the psychology of choice. *Science, 211*, 453-458.

Tversky, A. & Kahneman, D. (1984). Extensional versus intuitive reasoning: The conjunction fallacy in probability judgement. *Psychological Review, 90*, 293-315.

Ursin, H., Baade, E. & Levine, S. (1978). *Psychobiology of Stress*. New York: Academic Press.

Valentine, D. Williams, M. & Young, R. K. (1978). Age related factors in driving safety. Report prepared for National Highway Safety Administration, U.S. Dot.

Vando, A. (1974). The development of the R-A scale: A paper and pencil measure of pain tolerance. *Personality and Social Psychology Bulletin. 1*, 28-29.

Van der Colk, H. (1988). Risky behavior resulting from bounded rationality. *Ergonomics, 31*, 485-490.

Van der Molen, M.W., Somsen, R.J.M., Jennings, J.R., Nieuwboer, R. T. & Orlebeke, J.F. (1987). A psychophysiological investigation of cognitive-energetic relations in human information processing: a heart-rate/additive factors approach. *Acta Psychologica, 66*, 251-289.

Van der Molen, H.H. & Bötticher, A.M.T. (1988). A hierarchical model for traffic participants. *Ergonomics, 31*, 537-555.

Vanderveen, D. (1989). *Risk-Taking Stràtegy as a Function of Within-Individual Concordance Between Preferred Activities and their Perceived Riskiness*. B.A. Thesis. Queen's University, Kingston, Ontario, Canada.

Vanderwolf, C.H. & Robinson, T.E. (1981). Reticulo-cortical activity and behavior: A critique of the arousal theory and a new synthesis. *Behavior and Brain Sciences, 4*, 459-476.

Vari, A., Kemp, R., & Mumpower, J.L. (1991). Public concerns about LLRW facilitity siting. *Journal of Cross-Cultural Psychology, 22*, 83-102.
Vaughan, E. & Nordenstam, B. (1991). The perception of environmental risks among ethnically diverse groups. *Journal of Cross-Cultural Psychology, 22*, 29-60.
Vestewig, R.E. (1977). Extraversion and risk preference in portfolio theory. *Journal of Psychology*, *97*, 237-245.
Vincente, K. J., Thornton, D. C. & Moray, N. (1987). Spectral analysis of sinus arrhythmia: A measure of mental effort. *Human Factors*, *29*, 171-182.
Vlek, C. & Cvetkovich, G. (1989). *Social Decision Methodology for Technological Projects*. Dordrecht, Holland: Kluwer
Vlek, C.A.J. & Stallen. P.J. (1980). Rational and personal aspects of risk. *Acta Psychologica*, *45*, 273-300.
Vlek, C.A.J. & Stallen. P.J. (1981). Judging risks and benefits in the small and in the large. *Organizational Behavior and Human Performance*, *28*, 235-271.
Von Benda, H. & Hoyos, C. Graf (1983). Estimating hazards in traffic situations. *Accident Analysis and Prevention*, *15*, 1-9.
Von Neumann, J. & Morgenstern, O. (1947). *Theory of Games and Economic Behavior* (2nd Ed.). Princeton, New Jersey: Princeton University Press.
Von Winterfeldt, D., John, R.S. & Borcherding, K. (1981). Cognitive components of risk ratings. *Risk Analysis*, *1*, 277-287.
Von Winterfeldt, D. & Edwards, W. (1984). Patterns of conflict about risky technologies. *Risk Analysis*, *4*, 55-68.
Vroom, V.H. (1964). *Work and Motivation*. New York: Wiley.
Wagenaar, W.A., Hudson, P.T. & Reason, J.T. (1990). Cognitive failures and accidents. *Applied Cognitive Psychology*, *4*, 273-294.
Wagner, W. & Vierboom, C. (1991). Gefühlswelt junger Fahrer: Möglich keiten der Beeinflussung. *Unfall- und Sicherheitsforschung Strassen verkehr*, *84*. Bergisch Gladbach: Bundesanstalt für Straßenwesen.
Warbuton, D.M. (1979). Physiological Abstracts of Information Processing and Stress. In: V. Hamilton & Warburton, D.J. (Eds.). *Human Stress and Cognition. An Information Processing Approach*. Chichester: John Wiley & Sons.
Wasielewski, P. (1984). Speed as a Measure of Driver Risk: Observed Speed vs. Driver and Vehicle Characteristics. *Accident Analysis and Prevention, 16*, 89-102.

Weber, E.U. (1988). A Descriptive Measure of Risk. *Acta Psychologica, 69,* 185-203.
Weick, K.E. (1984). Samll Wins - Redefining the Scale of Social Problems. *American Psychologist, 39*, 40-49.
Weigold, M.F. & Schlenker, B.R. (1991). Accountability and Risk taking. *Personality and Social Psychology Bulletin, 17*, 25-29.
Weiner, B. (1991). Metaphors in Motivation and Attribution. *American Psychologist, 46*, 921-930.
Weiner, B. (1986). *An attributional theory of motivation and emotion.* New York: Springer Verlag
Wenninger, G. (1991). *Arbeitssicherheit und Gesundheit: psychologisches Grundwissen für betriebliche Sicherheitsexperten und Führungskräfte.* Heidelberg: Asanger.
Weiss, J.M., Glazer, H.I., Pohorecky, L.A., Brick, J. & Miller, N.E. (1975). Effects of chronic exposures to stressors on avoidance-escape behavior and on brain norepinephrine. *Psychosomatic Medicine, 37*, 522-534.
Weisz, J.R., & Stipek, D.J. (1982). Competence, contingency, and the development of perceived control. Human Development.
West, S.G. & Hepworth, J.T. (1989). An evaluation of Arizona`s July 1982 drunk driving law: effects on the city of Phoenix. *Journal of Applied Social Psychology, 19,* 1212-1237.
West, S.M. (1985). *Sex Differences in a Laboratory Test of the Risk Homeostasis theory*. B.A. Thesis. Queen's University, Kingston, Canada.
Whitaker, D.B. (1980). Personality traits as manifested in game behavior. *Perceptual and Motor Skills, 51*, 820.
White, R.W. (1959). Motivation reconsidered: The concept of competence. *Psychological Review. 66*, 297-333.
Wicker, F. W., Thorelli, I.M., Barron, W.L.III & Willis, A.C. (1981). Studies of mood and humour appreciation. *Motivation and Emotion, 5*, 47-59.
Wickens, C. D. (1979). Measures of workload, stress and secondary tasks. In: N. Moray (Ed.). *Mental Workload* (pp. 79-100). New York: Plenum Press.
Wickens, C. D. & Kramer, A. (1985). Engineering Psychology. *Annual Review of Psychology, 36*, 307-348.
Widom, C.S. (1977). A methodology for studying non-institutionalized psychopaths. *Journal of Consulting and Clinical Psychology, 45*, 674-683.
Widom, C.S. (1989). Does violence beget violence? A critical examination of the literature. *Psychological Bulletin, 106*, 3-28.

Widom, C.S. (1991). A tail on an untold tale: Response to:Biological and genetic contributors to violence-Widoms untold tale. *Psychological Bulletin, 109*, 130-132.
Wiederman, M.W. & Allgeier, E.R. (1992). Gender differences in mate selection criteria: Sociobiological or socioeconomic explanation? *Ethology and Sociobiology, 13*, 115-124.
Wiegner, P. (1982). Leistungsgrenzen bei der Führung von Motorrädern. *Zeitschrift für Verkehrssicherheit, 28*, 117-126.
Wierwille, W. W. (1979). Physiological measures of aircrew mental workload. *Human Factors, 21*, 575-594.
Wierwille, W. W., Rahimi, M. & Casali, J. G. (1985). Evaluation of 16 measures of mental workload using a simulated flight task emphasizing mediation activity. *Human Factors, 27*, 489-502.
Wildavsky, A. (1979). No risk is the highest risk of all. *American Scientist, 67*, 32-37.
Wilde, G.J.S. (1976). Social interaction patterns in driver behavior: An introductory review. *Human Factors, 5*, 477-492.
Wilde, G.J.S. (1978). Theorie der Risikokompensation der Unfallverursachung und praktische Folgerungen für die Unfallverhütung. *Hefte zur Unfallheilkunde, 130*, 134-156.
Wilde, G.J.S. (1982). The Theory of Risk Homeostasis: Implications for Safety and Health. *Risk Analysis, 2*, 209-225.
Wilde, G.J.S. (1982b). Critical Issues in Risk Homeostasis Theory. *Risk Analysis, 2*, 249-258.
Wilde, G.J.S. (1984). On the choice of denominator for the calculation of accident rates. In: S. Yagar (Ed.). *Transport Risk Assessment*. Proceedings of a symposium on Risk in Transport. Waterloo Ont.: University fo Waterloo Press.
Wilde, G.J.S. (1985) The use of incentives for the promotion of accident-free driving. *Journal of Studies on Alcohol, 10*, 161-167.
Wilde, G.J.S. (1985b). Assumptions necessary and unnecessary to risk homeostasis. *Ergonomics, 28*, 1531-1538.
Wilde, G.J.S. (1986). Beyond the concept of risk homeostasis: Suggestions for research and application towards the prevention of accidents and lifestyle-related disease. *Accident Analysis and Prevention, 18*,377-401.
Wilde, G.J.S. (1988). Risk homeostasis theory and traffic accidents: propositions, deductions and discussion of dissension in recent reactions. *Ergonomics, 31*, 441-468.

Wilde, G.J.S. (1988b). Risk-taking in psychomotor and cognitive tasks as a function of probability of loss, skill and other person related variables. In: Rothengatter, T. & de Bruin, R. (Eds.) *Road User Behavior: Theory And Research*. Assen, Maastricht NL.: Van Gorcum.

Wilde, G.J.S. (1989). Accident countermeasures and behavioral compensation: The position of risk homeostasis theory. *Journal of Occupational Accidents*, *10*, 267-292.

Wilde, G.J.S. (1991). Economics and Accidents: A Commentary. *Journal of Applied Behavior Analysis, 24*, 81-84.

Wilde, G.J.S. (1992). Accident prevention through incentives for safety in industry and road traffic: an analysis of international experience. XIII Internationales Sommer Symposion GFS. 15-17.6. 1992

Wilde, G.J.S. (1993). Effects of mass media communications on health and safety habits: an overview of issues and evidence. *Addiction, 88*, 983-996.

Wilde, G.J.S. (1993 in press). Risk homeostasis theory and its promise for improved safety (pp. 9-24) and New techniques for determining subjects`risk-taking tendency in their task performance (pp. 119-134). In: R. Trimpop & G.J.S. Wilde (Eds.). *Challenges to Accident Prevention: The issue of risk compensation behaviour*. Groningen, The Netherlands: STYX Publications.

Wilde, G.J.S., Claxton-Oldfield, S.P. & Platenius, P.H. (1985). Risk homeostasis in an experimental context. In: L.Evans & R.C. Schwing (Eds). *Human behavior and traffic safety* (pp.119-142). New York: Plenum,.

Wilde, G.J.S. & Kunkel, E. (1984). Die begriffliche und empirische Problematik der Risikokompensation! *Zeitschrift für Verkehrssicherheit, 30*, 52-61.

Wilde, G.J.S. & Murdoch, P.A. (1982). Incentive systems for accident-free and violation-free driving in the general population. *Ergonomics*, *25*, 879-890.

Wilde, G.J.S., Trimpop, R.M. & Joly, R. (1989). The effects of various amounts of ethanol upon risk taking tendency and confidence in task performance. In: *Procedings of the 11th International Conference on Alcohol, Drugs, and Traffic Safety*

Wilde, G.J.S. (1991). Economics and Accidents: A Commentary. *Journal of Applied Behavior Analysis, 24,* 81-84.

Wilde, G.S.J. (1992). Accident Prevention through Incentives for Safety in Industry and Road Traffic: an Analysis of International Experience. In: Proceedings of the XIII. Internationales GfS-Sommer-Symposium. Gelsenkirchen, Germany.

Wilde, G.J.S. (1993b). Effects of mass media communications on health and safety habits: an overview of issues and evidence. *Addiction, 88,* 983-996.
Wildervanck, C., Mulder, G. & Michon, J.A. (1978) Mapping mental load in car driving. *Ergonomics*, *21*, 225 - 229.
Williams, J.C. (1988). A Data-based method for assessing and reducing Human Error to improve operational performance. In Proceedings of IEEE 4th. Conference on Human Factors in Power Plants (pp. 436-450). Monterey, California.
Williams, J.C. (1992). Toward an Improved Evaluation Analysis Tool for Users of HEART. *International Conference on Hazard Identification and Risk Analysis, Human Factors and Human Reliability in Process Safety* (pp. 261-280). New York: American Institute of Chemical Engineers.
Wilson, E.O. (1975). *Sociobiology*. Cambridge:Harvard University Press.
Wilson, M. & Daly, M. (1985). Competitiveness, risk taking, and violence: the young male syndrome. *Ethology and Sociobiology*,*6*,59-73.
Winterfeldt, D. & Edwards, W. (1984). Decision analysis and behavioral research. Cambridge: Cambridge University Press
Wohlwill, J.F. (1984). What are sensation seekers seeking? *Behavior and Brain Science*, *7*, 453.
Woodworth, R.S. (1918). *Dynamic Psychology*. New York:Columbia University Press.
Wright, G. (1984). *Behavioral Decision Theory: An Introduction*. Harmondsworth, U.K.:Penguin.
Wuebker, L.J. (1986). Safety Locus of Control as a predictor of industrial accidents. *Journal of Business and Psychology*, *1*, 5-18.
Wundt, W. (1874). *Grundzüge der Physiologischen Psychologie* (1st ed.). Leipzig: Engelmann.
Yates, J.F. (1990). *Judgement and Decisionmaking*. Englewood Cliffs, NJ: Prentice Hall.
Yates, J.F. (Ed.) (1992). *Risk Taking Behavior*. Chichester: John Wiley & Sons.
Yates, J.F. & Stone, E.R. (1992). The Risk Construct. In: J.F. Yates (Ed.). *Risk Taking Behavior*. Chichester, UK: John Wiley & Sons.
Yerkes, R. M. & Dodson, J. D. (1908). The relation of strength of stimulus to rapidity of habit formation. *Journal of Comparative and Neurological Psychology*, *18*, 459-482.
Zajonc, R. (1980). Feeling and thinking: Preferences need no Inferences. *American Psychologist*, *35*, 151-175.

Zajonc, R. (1984). On the primacy of affect. *American Psychologist, 39*, 117-123.
Zaleski, Z. (1984). Sensation seeking and risk-taking behavior. *Personality and Individual Differences*, *5*, 607-608.
Ziller, R.C. (1957). Vocational choice and utility for risk. *Journal of Counseling Psychology*, *4*, 61-64.
Zimolong, B. (1985). Hazard perception and risk estimation in accident causation. *Trends in Ergonomics/Human Factors II*, 463-470.
Zimolong, B. (1990). Fehler und Zuverlässigkeit. In: C. Graf Hoyos & B. Zimolong (Hrsg.). *Enzyklopädie der Psychologie*, Bd.D-III-2: *Ingenieurpsychologie* (pp. 313-345). Göttingen: Hogrefe.
Zimolong, B. (1992). Empirical Evaluation of THERP, SLIM and Ranking to Estimate HEPs. *Reliability Engineering and System Safety*, *35*, 1-11.
Zimolong, B. & Gstalter, H. (1984). Gefahrenkognition bei Fahrzeug-Fußgänger Konflikten. *Zeitschrift für Verkehrssicherheit, 30*, 62-66.
Zimolong, B. & Trimpop, R.M. (1992). *Psychologie der Arbeitssicherheit.* Heidelberg: Asanger Verlag.
Zimolong, B. & Trimpop, R. (1993). *Managing Human Reliability in Advanced Manufacturing Systems*. (Chapter 15) In: Salvendy & Karwowski: *Design of Work and Development of Personnel in Manufacturing Systems*. New York: John Wiley & Sons.
Zinbarg, R. & Revelle, W. (1989). Personality and conditioning: A test of four models. *Journal of Personality and Social Psychology*,*57*,301-314.
Zink, K.J. & Ritter, A. (1992). *Mit Qualitätszirkeln zu mehr Arbeitssicherheit.* Wiesbaden: Universum Verlagsanstalt.
Zuckerman, M. (1971). Dimensions of sensation-seeking. *Journal of Consulting and Clinical Psychology*, *36*, 45-52.
Zuckerman, M. (1979). *Sensation Seeking: Beyond the Optimal Level of Arousal*. Hillsdale: Lawrence Erlbaum.
Zuckerman, M. (1980). Sensation seeking and its biological correlates. *Psychological Bulletin*, *88*, 187-214.
Zuckerman, M. (Ed.) (1983). *Biological Bases of Sensation Seeking, Impulsivity, and Anxiety*. Hillsdale: Lawrence Erlbaum.
Zuckerman, M. (1983 b). Sensation seeking and sports. *Personality and Individual Differences*, *4*, 285-293.
Zuckerman, M. (1989). Personality in the third dimension: A psychobiological approach. *Personality and Individual Differences*, *10*, 391-418.
Zuckerman, M. (1984). Sensation Seeking: A comparative approach to a human trait. *Behavior and Brain Science*, *7*, 413-434.

Zuckerman, M. (1984 b). Experience and desire: a new format for sensation seeking scales. *Journal of Behavioral Assessment*, *6*, 101-114.
Zuckerman, M. (1990). The psychophysiology of sensation seeking. *Journal of Personality*, *58*, 313-345.
Zuckerman, M. (1991). Some dubious Premises in Research and Theory on Racial Differences: Scientific, Social, and Ethical Issues. *American Psychologist*, *45*, 1297-1303.
Zuckerman, M., Buchsbaum, M.S. & Murphy, D.L. (1980). Sensation seeking and its biological correlates. *Psychological Bulletin*, *88*, 187-214.
Zuckerman, M., Eysenck, S. & Eysenck, H.J. (1978). Sensation seeking in England and America: Cross-cultural, age, and sex comparisons. *Journal of Consulting and Clinical Psychology*, *46*, 139-149.
Zuckerman, M., Kolin, E.A., Price, L. Zoob, I. (1964). Development of a Sensation-Seeking Scale. *Journal of Consulting and Clinical Psychology*, *28*, 477-482.
Zuckerman, M., Kuhlman, D.M. & Camac, C. (1988). What lies beyond E and N? Factor analyses of scales believed to measure basic dimensions of personality. *Journal of Personality and Social Psychology*, *54*, 96-107.
Zuckerman, M., Kuhlman, D.M. Thornquist, M. & Kiers, H. (1992). Five or three robust questionnaire scale factors of personality without culture. *Personality and Individual Differences*, *12*, 929-941.
Zuckerman, M. & Neeb, M. (1980). Demographic influences in sensation seeking and expressions of sensation seeking in religion, smoking and driving habits. *Personality and Individual Differences*, *1*, 197-206.
Zuckerman, M., Schultz, D.P. & Hopkins, T.R. (1967). Sensation-seeking and volunteering for sensory deprivation and hypnosis experiments. *Journal of Consulting and Clinical Psychology*, *31*, 358-363.
Zuckerman, M., Simons, R.F. & Como, P.G. (1988). Sensation seeking and stimulus intensity as modulators of cortical, cardiovascular, and electrodermal response: a cross-modality study. *Personality and Individual Differences*, *9*, 361-372.
Zylman, R. (1972). Age is more important than alcohol in the collision - involvement of young and old drivers. *Journal of Traffic Safety Education*, *20*, 7-8.
Zylman, R. (1974). A critical evaluation of the literature of "alcohol involvement" in highway deaths. *Accident Analysis and Prevention*, *6*, 163-204.
Zylman, R. (1975). Drinking-driving and fatal crashes: A new perspective. *Journal of Alcohol and Drug Education*, *21*, 1-10.

Author Index

C

D

S

Subject Index

U

V

W

Y

Z

FSC
www.fsc.org
MIX
Papier | Fördert
gute Waldnutzung
FSC® C083411

Kolibri 360 GmbH
Ferdinand-Jühlke-Straße 7
99095 Erfurt, Deutschland
produktsicherheit@kolibri360.de